Circuits and Systems:
An Engineering Perspective

Circuits and Systems:
An Engineering Perspective

Johnny Fuller

CLANRYE
INTERNATIONAL
www.clanryeinternational.com

Clanrye International,
750 Third Avenue, 9ᵗʰ Floor,
New York, NY 10017, USA

ISBN: 978-1-64726-128-3

Cataloging-in-Publication Data

Circuits and systems : an engineering perspective / Johnny Fuller.
p. cm.
Includes bibliographical references and index.
ISBN: 978-1-64726-128-3
1. Electronic circuits. 2. Electronic systems. 3. Electronic apparatus and appliances.
4. Electronics. I. Fuller, Johnny.
TK7867 .C57 2022
621.381 5--dc23

For information on all Clanrye International publications
visit our website at www.clanryeinternational.com

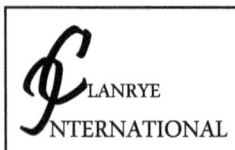

C LANRYE
INTERNATIONAL

TABLE OF CONTENTS

PREFACE

This book is a culmination of my many years of practice in this field. I attribute the success of this book to my support group. I would like to thank my parents who have showered me with unconditional love and support and my peers and professors for their constant guidance.

A complete electrical network in the form of a closed loop which gives a return path for electric current is known as an electrical circuit. There are various classifications of circuits such as on the basis of arrangement, type of current flowing through it, and the components. On the basis of arrangement, circuits are broadly divided to parallel circuits and series circuits. Circuits are classified as AC circuits and DC circuits, on the basis of the type of current which is flowing through it. System refers to the set of interacting entities which function together as a single unit. Study in the field of circuits and systems focuses on the analysis, theory and design of interconnected devices and components. The topics included in this book on circuits and systems are of utmost significance and bound to provide incredible insights to readers. It explores all the important aspects of these fields in the present day scenario. Scientists and students actively engaged in this field will find this book full of crucial and unexplored concepts.

The details of chapters are provided below for a progressive learning:

Chapter – Introduction

A circuit is a closed path around which a circulating current can flow. It is broadly classified into series circuits and parallel circuits. The chapter closely examines these types of circuits as well as the electrical circuit theory to provide an extensive understanding of the subject.

Chapter – Components of Circuits

Some of the major components of circuits are resistors, capacitors, inductors, voltage source, electric current, transistors, thyristors and diodes. They are connected by the conductive wires and traces through which electric current can flow. The topics elaborated in this chapter will help in gaining a better perspective about these components of circuits.

Chapter – Types of Circuits

Circuits can be classified into various types such as electrical circuits, AC circuits, DC circuits and electronic circuits. DC circuits and AC circuits involve the flow of DC current and AC current respectively. These types of circuits as well as the various subtypes of electronic circuits have been thoroughly discussed in this chapter.

Chapter – Circuit: Laws andTheorems

The major laws related to circuits are Kirchhoff's Voltage Law, Kirchhoff's Current Law, Ohm's Law, Joules Law of Heating, etc. The main theorems which are studied in relation to circuits are Thévenin's theorem, Norton's theorem, Reciprocity theorem and Millman's theorem. This chapter has been carefully written to provide an easy understanding of these laws and theorems related to circuits.

Chapter – Electronic Systems

The physical interconnection of components that gathers diverse amounts of information together is referred to as an electronic system. Some of the major focus areas related to electronic systems are system modelling, system analysis, system theory, complex adaptive system, etc. The chapter closely examines these key concepts of electronic systems to provide an extensive understanding of the subject.

Chapter – Control System and its Applications

Control systems are used to manage, direct, command or regulate the behavior of other systems or devices using control loops. Some of the different types of control systems are open loop control systems and feedback control systems. This chapter discusses in detail these concepts related to control systems as well as the applications of control theory in biomedical engineering.

Johnny Fuller

CHAPTER 1

Introduction

A circuit is a closed path around which a circulating current can flow. It is broadly classified into series circuits and parallel circuits. The chapter closely examines these types of circuits as well as the electrical circuit theory to provide an extensive understanding of the subject.

CIRCUIT

A circuit is the path that an electric current travels on, and a simple circuit contains three components necessary to have a functioning electric circuit, namely, a source of voltage, a conductive path, and a resistor.

Circuits are driven by flows. Flows are ubiquitous in nature and are often the result of spatial differences in potential energy. Water flows downriver due to changes in height, tornadoes swirl due to gentle temperature gradients, and sucrose flows from the leaves of trees to their distal parts due to differences in water potential. Even life itself is due to a clever hack by which living organisms serve as a conduit for the flow of solar energy. Perhaps then it is no surprise that electronic devices are driven by flows.

In a simple circuit, voltage flows through the conductive path to the resistor, which does some work. Resistors--things like light bulbs, speakers, and motors-and electric circuits power these devices to do the work that their makers wanted them to do.

River-dam Analogy

Flows and circuits can be illustrated via analogies of water, rivers, lakes and dams. If there are two lakes, with a trench dug to connect them, and both lakes have water at the same height $h_1 = h_2$, then the water in the trench will not flow in either direction. There isn't anything driving the system. A given parcel of water has the same potential energy and feels the same atmospheric pressure whether in one lake or the other, so there can be no net transport of water.

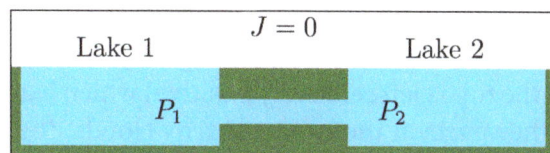

If this changes, and the height of the water in one lake is greater than the other, e.g. $h_2 > h_1$, then water will flow. Now, the pressure on the trench from the water in the higher lake is greater than the corresponding pressure from the lower lake, and hence water should flow out of the higher lake, through the trench, and into the lower lake. If it were possible to keep the water level in the two lakes constant, for instance by replacing the water that leaves the high lake, and removing the water that enters the low lake, then there would be a steady flow of water through the trench.

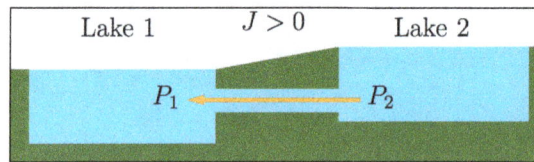

Moreover the flowspeed could be altered by altering the lakes and trench. If, for example, the trench is widened, or is built out of smooth materials, the flow should quicken, and if the trench is constricted, or filled with rough material and debris, the flow should slow down.

The simplest phenomenological relation consistent with this logic is,

$$\text{flow} = \frac{\text{push}}{\text{resistance}}.$$

The push can take many forms including, but not limited to, physical forces, gas and liquid pressures, or differences in potential energy.

In the case of the two lakes, the flow J is the transport of water from one place to another in response to the push, a difference in pressure Δp. The flow can be altered by the characteristics of the flow path, which can be rolled into a descriptive number called the hydraulic resistance, R_H For the water flowing through the trench,

$$J = \frac{\Delta P}{RH}.$$

This kind of linear relationship between push and flow is found time and again in many different situations. Often, this kind of relation can get a working model off the ground in systems where very little is known about the detailed workings, and where one might otherwise be paralyzed by details they don't yet understand. However, with a minimal model that can successfully relate measurable bulk properties (like the flow, the push, and the resistance), everyone is in a better position to question the origin of these quantities and build toward a more systematic understanding. Such is the spirit in which the study of electrical circuits began.

Phenomenological Relation - Ohm's Law

Long ago, people noticed that lightning, i.e. charged matter, can move from one place to another en masse. The reason for this is that clouds build up a large asymmetry in charge (i.e. electrons accumulate at the bottom, and the top is left relatively positive) which leaves parts of the cloud highly charged in comparison to other parts of the cloud, nearby clouds, the ground, or even airplanes. This asymmetry creates a large difference in electric potential between the charged region of cloud and other objects. In the case of cloud to ground lightning, the negatively charged bottom region of a cloud has a large electric potential relative to the Earth (which has a net charge approximately equal to zero), on the order of 10^8 volts. This is a situation abhorred by nature, and works to relax these gaps in electric potential by flowing charge to balance out the asymmetries.

In the 18th century, people began to master the production of batteries, which are devices that can maintain significant gaps of electric potential between two points in space.

When the two points are brought into contact, e.g. by connecting them with a wire, the circuit is "closed" and the battery does work to flow a current, i.e. move charged particles from one end of the wire toward the other. In so doing, the internal potential of the battery is relaxed, just as is the case with the cloud and ground in a lightning strike. Discharging a battery through a loop of wire isn't all that useful, but a battery can be used to drive current through an electrical circuit such as a lightbulb, a circuit-board, a ceiling fan, or a sound system.

When electric potential, i.e. voltage, of a given magnitude is maintained over the terminals of a device, a current will flow through the device. The strength of this current has a linear relation $V \sim I$ to the applied voltage. The device terminal connected to the positive end of a V volt battery is kept at high potential (V volts relative to the negative end), and the terminal connected to the negative end is kept at low potential ($-V$ volts relative to the positive end).). The greater the gap in electric potential, the greater the current that will flow through the device.

For every device, the ratio V / I is given by a parameter, called the "resistance," commonly symbolized as Ω If the same voltage is applied to two devices, and one flows half as much current as the other, the device with half as much current is said to be twice as resistive as the other. This resistance is perfectly analogous to the hydraulic resistance from the thought experiment of changing the flow of water in the trench by smoothing its surface ((increasing flow, decreasing R_H or filling it with heavy debris ((decreasing flow, increasing R_H). This relation can be expressed as,

$$V = IR,$$

the so-called Ohm's law of circuits, yet another manifestation of the phenomenological push-flow relationship described above.

Units of Resistance

It's now possible to make some observations about the nature of R, the circuit resistance. Rearranging Ohm's law shows $R = V / I$, i.e. the resistance of a circuit is the cost in voltage to achieve a current of magnitude I through that circuit. The physical units of the volt are energy per charge, i.e. joules per coulomb (J / C) in the SI system. The unit of current is simply charge per unit time, i.e. coulomb per second (C / s) in the SI system, also called amperes, written as A or "amp". Thus, the units of resistance are Js / C^2, commonly called the Ohm Ω.

Remain in Light

Suppose a floor lamp which is rated at a resistance of 5Ω requires a current of 70 Amps to properly function. What must be the difference in voltage between the prongs of the plug, as provided by the wall?

Because the current that is drawn from the wall obeys Ohm's relation, we can say that

$$V_{\text{plug}} = IR_{\text{lamp}}, \text{and thus } V_{\text{plug}} = 350 \text{ volts.}$$

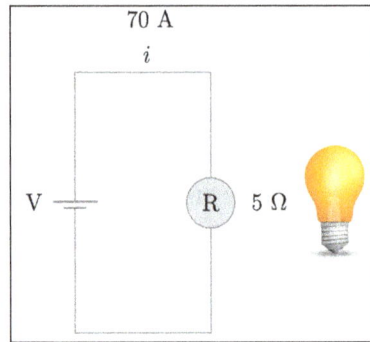

Resistors in Series

For any simple system, finding V, I, or R is straightforward when given the other two factors, but it gets more complicated when a power source drives multiple devices in series. Series means several devices connected end-to-end, with the positive terminal of one device connected to the negative device of the next, just like a set of Christmas lights. Because the devices flow into one another, and charge is conserved, any current that flows into the first device must flow out from the last device, i.e. the current through every device is the same. Devices in series are like water floating down a river: the river can twist, turn, contract, and expand, but the amount of water flowing by any given cross section per unit time must be the same at all points along the river i.e $v_1 A_1 = v_2 A_2$ If this were not so, water would build-up at points along the river and would overflow the banks.

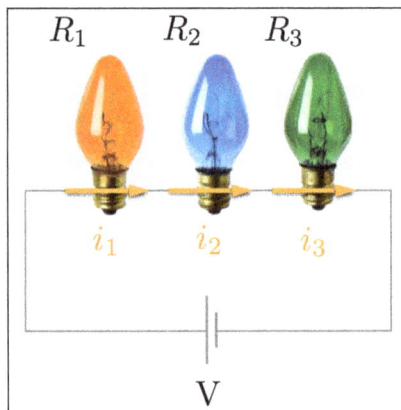

Thus in the circuit above, $i_1 = i_2 = i_3$ or since each resistor obeys Ohm's law,

$$I = \frac{V_1}{R_1} = \frac{V_2}{R_2} = \frac{V_3}{R_3}.$$

Now, the left side of the orange bulb is connected to the positive terminal of the battery, and the right side of the green bulb is connected to the negative terminal of the battery, which means that the sum of the voltage drops across the three resistors is equal in magnitude to the voltage drop across the battery, i.e.

$$V_{battery} = V_1 + V_2 + V_3.$$

This is the physical principle.

Hence,

$$V_{\text{battery}} = V_1 + V_2 + V_3$$
$$= IR_1 + IR_2 + IR_3$$
$$= I(R_1 + R_2 + R_3)$$
$$= IR_{\text{eff}}.$$

Therefore, a circuit consisting of three bulbs in series is equivalent to a single bulb with resistance equal to the sum of the individual resistances. This proves the general result for resistors in series.

Resistors in Series

The effective resistance of resistors $R_1, \ldots, R_N$ in series is equal to

$$R_{\text{eff}} = \sum_i R_i.$$

While arranging circuit elements in series has some attractive features like uniform current, ease of introducing new batteries, etc., there are major drawbacks to arranging circuit elements in series. For one, introducing any new devices decreases the current flowing through the circuit, and thus reduces the power output of every single device. If multiple devices are connected in series, for instance, your oven, your computer, and your reading lamp, dimming your reading lamp (by increasing its resistance) means less current to your oven and computer. Another is that if one element in the circuit, your TV for instance, breaks, the entire circuit will also break because the electric potential gap is no longer maintained across any device. This is inconvenient for building durable circuitry, where we'd like device failures to be independent of one another.

Some of these drawbacks can be avoided in parallel circuit architecture.

Resistors in Parallel

In parallel arrangements, each circuit element is connected to the terminals of the battery independent of the other circuit elements. Because their terminals are each held at the potentials of the battery terminals, the voltage across each device is equal to the voltage across the battery itself. If one of the devices experiences a failure (i.e. the path for current breaks in a given device), the other devices continue to function unabated. Again, we wish to know what happens when a battery drives several devices in parallel, i.e. what is the effective resistance of connecting devices in parallel? Consider the diagram below, depicting a set of resistors in parallel, connected to a battery of voltage

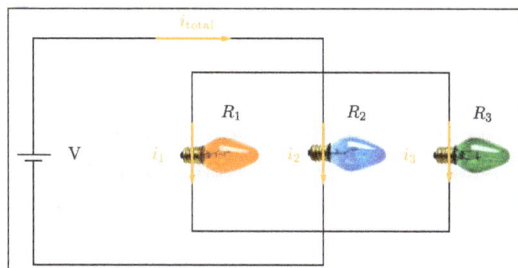

The total current coming out of the i_{total} splits into three currents i_1, i_2, and i_3, i.e.

$$i_{total} = i_1 + i_2 + i_3.$$

This is the physical principle.

As each element follows Ohm's law, and each element has the same voltage drop $V = I_1 R_1 = I_2 R_2 = I_3 R_3$ it follows that $I_i = \dfrac{V}{R_i}$. Also, because the total current is conserved:

$$I_{total} = \sum_i I_i$$

$$= \sum_i \frac{V_{battery}}{R_i}$$

$$= V_{battery} \left(\frac{1}{R_i} + \frac{1}{R_2} + \frac{1}{R_3} \right)$$

$$= \frac{V_{battery}}{R_{eff}}.$$

Thus, the effective resistance of resistors in parallel is given by the inverse of the sum of the inverse resistances.

Parallel Resistance

The effective resistance of a set of resistors $R_1, \ldots, R_N$ in parallel is given by,

$$R_{eff} = \left(\sum_i R_i^{-1} \right)^{-1}.$$

Infinite Ladder of Resistors

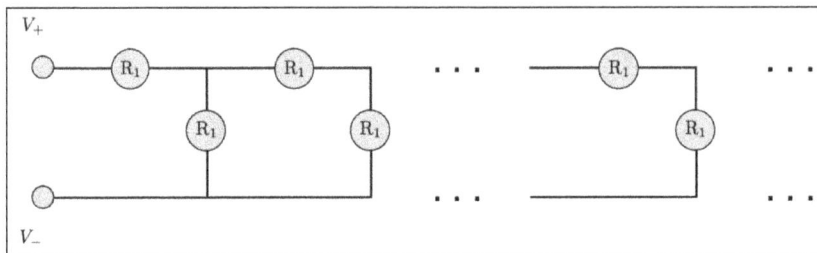

Calculate the resistance between points V_+ and V_- in the diagram above.

Studying the circuit diagram, we see that starting from point V_+, the current encounters a single resistor R_1 in series with a branch that has another resistor R_1 in parallel with an infinite ladder. In principle, we can write down new equations every time the circuit makes a new branch, but that will lead to a rather large system of relations to solve. It might be profitable to think about the remainder of the circuit as a black box device of some effective resistance. If we look at the circuit within the black (gray in the diagram below) box, we notice that it is an exact copy of the overall circuit. Of course, it is missing the first bit of circuit that falls outside the gray box, but this is of no consequence as the ladder is infinite. The difference is analogous to subtracting 1 from ∞, and there is no difference between ∞, and $\infty - 1$.

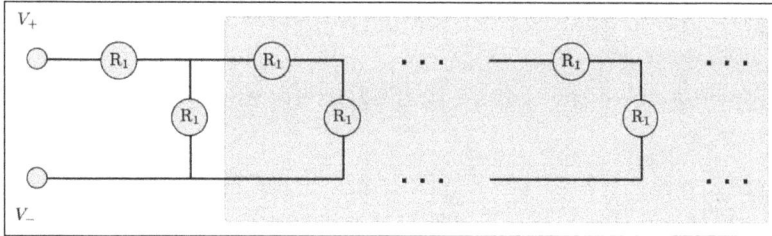

If we call the resistance of the gray box R_{ladder}, we have the following expression for the overall resistance:

$$R_{ladder} = R_1 + \left(\frac{1}{R_1} + \frac{1}{R_{ladder}} \right)^{-1}.$$

Multiplying through by $\left(\frac{1}{R_1} + \frac{1}{R_{ladder}} \right)$, we have

$$\frac{R_{ladder}}{R_1} + 1 = \frac{R_1}{R_{ladder}} + 2.$$

Simplifying, we find the quadratic equation,

$$R_{ladder}^2 - R_1 R_{ladder} - R_1^2 = 0,$$

which yields the solution $R_{ladder} = R_1 \frac{1}{2}(1 + \sqrt{5}) = R_1 \phi$, where ϕ is the golden ratio.

Kirchhoff's Current Conservation Law

In the parallel resistor discussion above, at the point where the main wire splits into three, the total current is conserved. This principle holds in general, whenever a collection of wires meets at a node. Charge is conserved and all currents must end up somewhere, and therefore the sum of the incoming currents minus the sum of the outgoing currents must equal zero. This is one of the main tools in circuit analysis and is commonly known as Kirchhoff's current law.

Kirchhoff's Current Law

All incoming currents to a junction of wires must exit the junction:

$$\sum_{in} I = \sum_{out} I,$$

or if we use the convention that incoming and outgoing currents have opposite signs,

$$\sum_i I_i = 0.$$

Kirchhoff's Loop Law

In the series resistors discussion, the voltage across the battery was equal to the sum of the voltages across the other circuit elements. Further, if an electron moves down a voltage drop V, the electron will pick up the kinetic energy $q_e V$. Similarly, to bring an electron up a gradient of voltage V, the electron will lose the energy $q_e V$. Assuming that electrons start from the battery at rest, the energy gained by dropping down the voltage of the battery must equal precisely the energy lost by traversing the resistors.

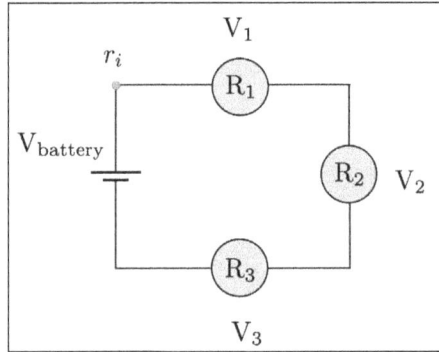

If this were not the case, starting at some point r_i in the loop (yellow dot in the above figure), then traveling around the loop, thus changing the potential by the amount $\Delta V = V_{battery} - V_1 - V_2 - V_3$, would result in ending up back at r_i at a higher potential than the journey began with, i.e. point r_i would have the electric potential ΔV relative to itself. Thus, the voltage of the battery and the voltages across the resistors must have opposite orientations, and the voltage around any closed loop must be zero. This is known as Kirchhoff's loop law.

SERIES CIRCUITS

A series circuit is characterised as having only a single path for current flow.

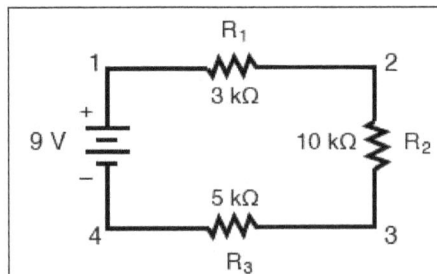

The first principle to understand about series circuits is as follows:

The amount of current in a series circuit is the same through any component in the circuit.

This is because there is only one path for current flow in a series circuit. Because electric charge flows through conductors like marbles in a tube, the rate of flow (marble speed) at any point in the circuit (tube) at any specific point in time must be equal.

Using Ohm's Law in Series Circuits

From the way that the 9-volt battery is arranged, we can tell that the current in this circuit will flow in a clockwise direction, from point 1 to 2 to 3 to 4 and back to 1. However, we have one source of voltage and three resistances. How do we use Ohm's Law here?

An important caveat to Ohm's Law is that all quantities (voltage, current, resistance, and power) must relate to each other in terms of the same two points in a circuit. We can see this concept in action in the single resistor circuit example below.

Using Ohm's Law in a Simple, Single Resistor Circuit

With a single-battery, single-resistor circuit, we could easily calculate any quantity because they all applied to the same two points in the circuit:

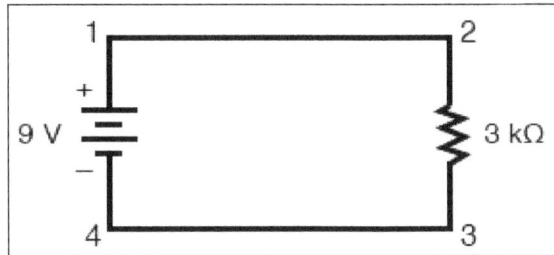

$$1 = \frac{E}{R}$$

$$1 = \frac{9 \ volts}{3 \ k\Omega} = 3 \ \text{mA}$$

Since points 1 and 2 are connected together with the wire of negligible resistance, as are points 3 and 4, we can say that point 1 is electrically common to point 2, and that point 3 is electrically common to point 4. Since we know we have 9 volts of electromotive force between points 1 and 4 (directly across the battery), and since point 2 is common to point 1 and point 3 common to point 4, we must also have 9 volts between points 2 and 3 (directly across the resistor).

Therefore, we can apply Ohm's Law (I = E/R) to the current through the resistor, because we know the voltage (E) across the resistor and the resistance (R) of that resistor. All terms (E, I, R) apply to the same two points in the circuit, to that same resistor, so we can use the Ohm's Law formula with no reservation.

Using Ohm's Law in Circuits with Multiple Resistors

In circuits containing more than one resistor, we must be careful in how we apply Ohm's Law. In the three-resistor example circuit below, we know that we have 9 volts between points 1 and 4, which is the amount of electromotive force driving the current through the series combination of R_1, R_2, and R_3. However, we cannot take the value of 9 volts and divide it by 3k, 10k or 5k Ω to try to find a current value, because we don't know how much voltage is across any one of those resistors, individually.

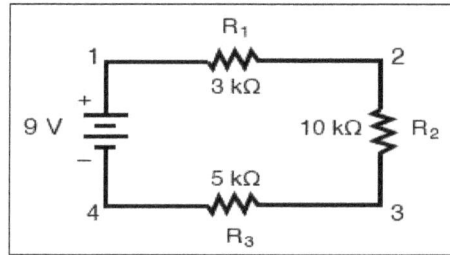

The figure of 9 volts is a *total* quantity for the whole circuit, whereas the figures of 3k, 10k, and 5k Ω are *individual* quantities for individual resistors. If we were to plug a figure for total voltage into an Ohm's Law equation with a figure for individual resistance, the result would not relate accurately to any quantity in the real circuit.

For R_1, Ohm's Law will relate the amount of voltage across R_1 with the current through R_1, given R_1's resistance, 3kΩ:

$$I_{R1} = \frac{E_{R1}}{3\ k\Omega} \quad E_{R1} = 1_{R1}\left(3k\ \Omega\right)$$

But, since we don't know the voltage across R_1 (only the total voltage supplied by the battery across the three-resistor series combination) and we don't know the current through R_1, we can't do any calculations with either formula. The same goes for R_2 and R_3: we can apply the Ohm's Law equations if and only if all terms are representative of their respective quantities between the same two points in the circuit.

So what can we do? We know the voltage of the source (9 volts) applied across the series combination of R_1, R_2, and R_3, and we know the resistance of each resistor, but since those quantities aren't in the same context, we can't use Ohm's Law to determine the circuit current. If only we knew what the *total* resistance was for the circuit: then we could calculate the *total* current with our figure for *total* voltage (I=E/R).

Combining Multiple Resistors into an Equivalent Total Resistor

This brings us to the second principle of series circuits:

The total resistance of any series circuit is equal to the sum of the individual resistances.

This should make intuitive sense: the more resistors in series that the current must flow through, the more difficult it will be for the current to flow.

In the example problem, we had a 3 kΩ, 10 kΩ, and 5 kΩ resistors in series, giving us a total resistance of 18 kΩ:

$$R_{total} = R_1 + R_2 + R_3$$
$$R_{total} = 3\ k\Omega + 10k\Omega + 5\ k\Omega$$
$$R_{total} = 18\ k\Omega$$

In essence, we've calculated the equivalent resistance of R_1, R_2, and R_3 combined. Knowing this, we could redraw the circuit with a single equivalent resistor representing the series combination of R_1, R_2, and R_3:

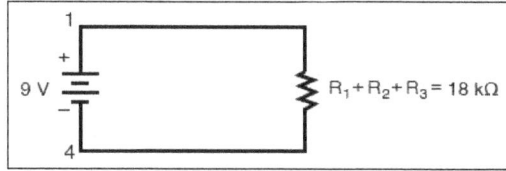

Calculating Circuit Current using Ohm's Law

Now we have all the necessary information to calculate circuit current because we have the voltage between points 1 and 4 (9 volts) and the resistance between points 1 and 4 (18 kΩ):

$$I_{total} = \frac{E_{total}}{R_{total}}$$

$$I_{total} = \frac{9\,volts}{18\ k\Omega} = 500\ \mu A$$

Calculating Component Voltages using Ohm's Law

Knowing that current is equal through all components of a series circuit (and we just determined the current through the battery), we can go back to our original circuit schematic and note the current through each component:

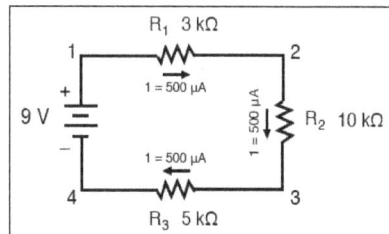

Now that we know the amount of current through each resistor, we can use Ohm's Law to determine the voltage drop across each one (applying Ohm's Law in its proper context):

$$E_{R1} = I_{R1}\ R_1 \quad E_{R2} = I_{R2}\ R_2 \quad E_{R3} = I_{R3}\ R_3$$

$$E_{R1} = (500\mu\ A)(3\,k\Omega) = 1.5\ V$$

$$E_{R2} = (500\mu\ A)(10\,k\Omega) = 5\,V$$

$$E_{R3} = (500\ \mu A)(5\,k\Omega) = 2.5\,V$$

Notice the voltage drops across each resistor, and how the sum of the voltage drops (1.5 + 5 + 2.5) is equal to the battery (supply) voltage: 9 volts.

This is the third principle of series circuits:

The supply voltage in a series circuit is equal to the sum of the individual voltage drops.

Analyzing Simple Series Circuits with the "Table Method" and Ohm's Law

However, the method we just used to analyze this simple series circuit can be streamlined for better understanding. By using a table to list all voltages, currents, and resistance in the circuit, it becomes very easy to see which of those quantities can be properly related in any Ohm's Law equation.

	R_1	R_2	R_3	Total	
E					Volts
I					Amps
R					Ohms

Ohm's Law   Ohm's Law   Ohm's Law   Ohm's Law

The rule with such a table is to apply Ohm's Law only to the values within each vertical column. For instance, E_{R_1} only with I_{R_1} and R_1; E_{R_2} only with I_{R_2} and R_2; etc. You begin your analysis by filling in those elements of the table that are given to you from the beginning:

	R_1	R_2	R_3	Total	
E				**9**	Volts
I					Amps
R	**3k**	**10k**	**5k**		Ohms

As you can see from the arrangement of the data, we can't apply the 9 volts of ET (total voltage) to any of the resistances (R_1, R_2, or R_3) in any Ohm's Law formula because they're in different columns. The 9 volts of battery voltage is *not* applied directly across R_1, R_2, or R_3. However, we can use our "rules" of series circuits to fill in blank spots on a horizontal row. In this case, we can use the series rule of resistances to determine a total resistance from the *sum* of individual resistances:

	R_1	R_2	R_3	Total	
E				9	Volts
I					Amps
R	3k	10k	5k	**18k**	Ohms ←

Rule of series circuits

$$R_T = R_1 + R_2 + R_3$$

Now, with a value for total resistance inserted into the rightmost ("Total") column, we can apply Ohm's Law of I=E/R to total voltage and total resistance to arrive at a total current of 500 µA:

	R_1	R_2	R_3	Total	
E				9	Volts
I				**500µ**	Amps
R	3k	10k	5k	18k	Ohms

Ohm's Law

Then, knowing that the current is shared equally by all components of a series circuit (another "rule" of series circuits), we can fill in the currents for each resistor from the current figure just calculated:

Finally, we can use Ohm's Law to determine the voltage drop across each resistor, one column at a time.

	R_1	R_2	R_3	Total	
E	**1.5**	**5**	**2.5**	9	Volts
I	500µ	500µ	500µ	500µ	Amps
R	3k	10k	5k	18k	Ohms

↑ ↑ ↑
Ohm's Ohm's Ohm's
Law Law Law

Verifying Calculations with Computer Analysis

Just for fun, we can use a computer to analyze this very same circuit automatically. It will be a good way to verify our calculations and also become more familiar with computer analysis. First, we have to describe the circuit to the computer in a format recognizable by the software. The SPICE program we'll be using requires that all electrically unique points in a circuit be numbered, and component placement is understood by which of those numbered points, or "nodes," they share. For clarity, I numbered the four corners of our example circuit 1 through 4. SPICE, however, demands that there be a node zero somewhere in the circuit, so I'll redraw the circuit, changing the numbering scheme slightly:

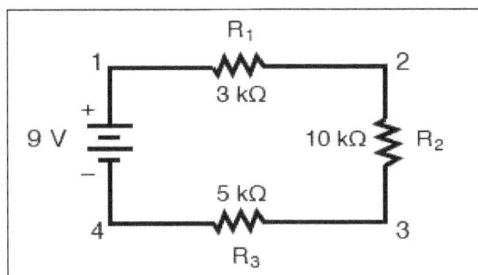

All we have done here is re-numbered the lower-left corner of the circuit 0 instead of 4. Now, I can enter several lines of text into a computer file describing the circuit in terms SPICE will understand, complete with a couple of extra lines of code directing the program to display voltage and current data for our viewing pleasure. This computer file is known as the *netlist* in SPICE terminology:

```
series circuit

v1 1 0

r1 1 2 3k

r2 2 3 10k

r3 3 0 5k

.dc v1 9 9 1
```

```
.print dc v(1,2) v(2,3) v(3,0)

.end
```

Now, all we have to do is run the SPICE program to process the netlist and output the results:

v1	v(1,2)	v(2,3)	v(3)	i(v1)
9.000E+00	1.500E+00	5.000E+00	2.500E+00	-5.000E-04

This printout is telling us the battery voltage is 9 volts, and the voltage drops across R_1, R_2, and R_3 are 1.5 volts, 5 volts, and 2.5 volts, respectively. Voltage drops across any component in SPICE are referenced by the node numbers the component lies between, so v(1,2) is referencing the voltage between nodes 1 and 2 in the circuit, which are the points between which R_1 is located.

The order of node numbers is important: when SPICE outputs a figure for v(1,2), it regards the polarity the same way as if we were holding a voltmeter with the red test lead on node 1 and the black test lead on node 2. We also have a display showing current (albeit with a negative value) at 0.5 milliamps or 500 microamps. So our mathematical analysis has been vindicated by the computer. This figure appears as a negative number in the SPICE analysis, due to a quirk in the way SPICE handles current calculations.

PARALLEL CIRCUITS

A parallel circuit is characterized by having the same voltage across every component in the circuit.

Voltage in Parallel Circuits

The first principle to understand about parallel circuits is that the voltage is equal across all components in the circuit. This is because there are only two sets of electrically common points in a parallel circuit, and the voltage measured between sets of common points must always be the same at any given time.

Therefore, in the above circuit, the voltage across R_1 is equal to the voltage across R_2 which is equal to the voltage across R_3 which is equal to the voltage across the battery.

This equality of voltages can be represented in another table for our starting values:

	R_1	R_2	R_3	Total	
E	2	6	1	9	Volts
I	20m	20m	20m	20m	Amps
R	100	300	50	450	Ohms

Ohm's Law Applications for Simple Parallel Circuits

Just as in the case of series circuits, the same caveat for Ohm's Law applies: values for voltage, current, and resistance must be in the same context in order for the calculations to work correctly.

However, in the above example circuit, we can immediately apply Ohm's Law to each resistor to find its current because we know the voltage across each resistor (9 volts) and the resistance of each resistor:

$$I_{R1} = \frac{E_{R1}}{R_1} \quad I_{R2} = \frac{E_{R2}}{R_2} \quad I_{R3} = \frac{E_{R3}}{R_3}$$

$$I_{R1} = \frac{9V}{10\,k\Omega} = 0.9\,mA$$

$$I_{R2} = \frac{9V}{2\,k\Omega} = 4.5\,mA$$

$$I_{R3} = \frac{9V}{1k\Omega} = 9\,mA$$

	R_1	R_2	R_3	Total	
E	9	9	9	9	Volts
I	0.9m	4.5m	9m		Amps
R	10k	2k	1k		Ohms

	↑	↑	↑
	Ohm's Law	Ohm's Law	Ohm's Law

At this point, we still don't know what the total current or total resistance for this parallel circuit is, so we can't apply Ohm's Law to the rightmost ("Total") column. However, if we think carefully about what is happening, it should become apparent that the total current must equal the sum of all individual resistor ("branch") currents:

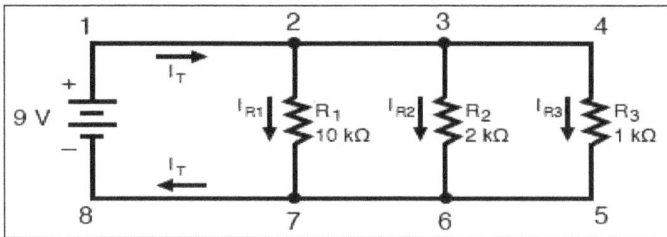

As the total current exits the positive (+) battery terminal at point 1 and travels through the circuit, some of the flow splits off at point 2 to go through R_1, some more splits off at point 3 to go through R_2, and the remainder goes through R_3. Like a river branching into several smaller streams, the combined flow rates of all streams must equal the flow rate of the whole river.

The same thing is encountered where the currents through R_1, R_2, and R_3 join to flow back to the negative terminal of the battery (-) toward point 8: the flow of current from point 7 to point 8 must equal the sum of the (branch) currents through R_1, R_2, and R_3.

This is the second principle of parallel circuits: the total circuit current is equal to the sum of the individual branch currents.

Using this principle, we can fill in the IT spot on our table with the sum of I_{R1}, I_{R2}, and I_{R3}:

	R_1	R_2	R_3	Total	
E	9	9	9	9	Volts
I	0.9m	4.5m	9m	**14.4m**	Amps
R	10k	2k	1k		Ohms

Rule of parallel circuits

$I_{total} = I_1 + I_2 + I_3$

How to Calculate Total Resistance in Parallel Circuits

Finally, applying Ohm's Law to the rightmost ("Total") column, we can calculate the total circuit resistance:

	R_1	R_2	R_3	Total	
E	9	9	9	9	Volts
I	0.9m	4.5m	9m	14.4m	Amps
R	10k	2k	1k	**625**	Ohms

$$R_{total} = \frac{E_{total}}{I_{total}} = \frac{9\ V}{14.4\ mA} = 625\ \Omega \quad \text{Ohm's Law}$$

The Equation for Resistance in Parallel Circuits

Please note something very important here. The total circuit resistance is only 625 Ω: *less* than any one of the individual resistors. In the series circuit, where the total resistance was the sum of the individual resistances, the total was bound to be *greater* than any one of the resistors individually.

Here in the parallel circuit, however, the opposite is true: we say that the individual resistances diminishrather than add to make the total.

This principle completes our triad of "rules" for parallel circuits, just as series circuits were found to have three rules for voltage, current, and resistance.

Mathematically, the relationship between total resistance and individual resistances in a parallel circuit looks like this:

$$R_{total} = \frac{1}{\dfrac{1}{R_1} + \dfrac{1}{R_2} + \dfrac{1}{R_3}}$$

How to Alter Parallel Circuit Numbering Schemes for SPICE

The same basic form of the equation works for *any* number of resistors connected together in parallel, just add as many 1/R terms on the denominator of the fraction as needed to accommodate all parallel resistors in the circuit.

Just as with the series circuit, we can use computer analysis to double-check our calculations.

First, of course, we have to describe our example circuit to the computer in terms it can understand. I'll start by re-drawing the circuit:

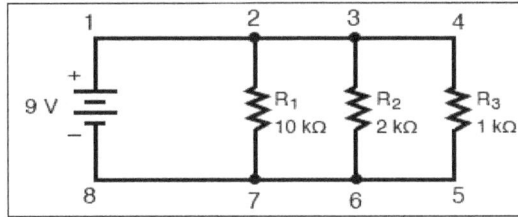

Once again, we find that the original numbering scheme used to identify points in the circuit will have to be altered for the benefit of SPICE.

In SPICE, all electrically common points must share identical node numbers. This is how SPICE knows what's connected to what and how.

In a simple parallel circuit, all points are electrically common in one of two sets of points. For our example circuit, the wire connecting the tops of all the components will have one node number and the wire connecting the bottoms of the components will have the other.

Staying true to the convention of including zero as a node number, I choose the numbers 0 and 1:

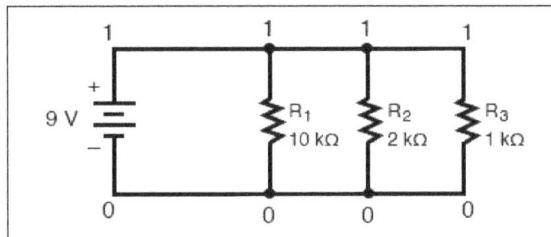

An example like this makes the rationale of node numbers in SPICE fairly clear to understand. By having all components share common sets of numbers, the computer "knows" they're all connected in parallel with each other.

In order to display branch currents in SPICE, we need to insert zero-voltage sources in line (in series) with each resistor, and then reference our current measurements to those sources.

For whatever reason, the creators of the SPICE program made it so that current could only be calculated *through* a voltage source. This is a somewhat annoying demand of the SPICE simulation program. With each of these "dummy" voltage sources added, some new node numbers must be created to connect them to their respective branch resistors:

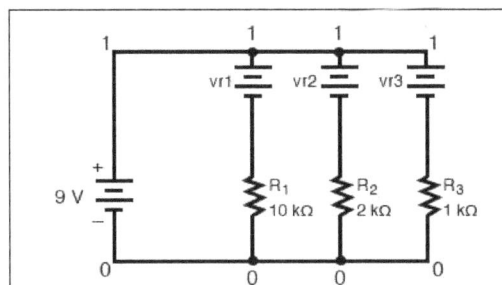

How to Verify Computer Analysis Results

The dummy voltage sources are all set at 0 volts so as to have no impact on the operation of the circuit.

The circuit description file, or *netlist*, looks like this:

```
Parallel circuit

v1 1 0

r1 2 0 10k

r2 3 0 2k

r3 4 0 1k

vr1 1 2 dc 0

vr2 1 3 dc 0

vr3 1 4 dc 0

.dc v1 9 9 1

.print dc v(2,0) v(3,0) v(4,0)

.print dc i(vr1) i(vr2) i(vr3)

.end
```

Running the computer analysis, we get these results (I've annotated the printout with descriptive labels):

v1	v(2)	v(3)	v(4)
9.000E+00	9.000E+00	9.000E+00	9.000E+00
Battery Voltage	R1 voltage	R2 voltage	R3 voltage

v1	i(vr1)	i(vr2)	i(vr3)
9.000E+00	9.000E-04	4.500E-03	9.000E-03
Battery Voltage	R1 current	R2 current	R3 current

These values do indeed match those calculated through Ohm's Law earlier: 0.9 mA for I_{R1}, 4.5 mA for I_{R2}, and 9 mA for I_{R3}. Being connected in parallel, of course, all resistors have the same voltage dropped across them (9 volts, same as the battery).

Three Rules of Parallel Circuits

In summary, a parallel circuit is defined as one where all components are connected between the same set of electrically common points. Another way of saying this is that all components are connected across each other's terminals.

From this definition, three rules of parallel circuits follow:

- All components share the same voltage.

- Resistances diminish to equal a smaller, total resistance.

- Branch currents add to equal a larger, total current.

Just as in the case of series circuits, all of these rules find root in the definition of a parallel circuit.

ELECTRICAL CIRCUIT THEORY

There are few theorems that can be applied to find the solution of electrical networks by simplifying the network itself or it can be used to calculate their analytical solution easily. The electrical circuit theorems can also be applied to A.C systems, with only one difference: replacing the ohmic resistance of the D.C system with impedance.

There are two general approaches to network analysis:

Direct Method

In this method, the network is left in its original form while determining it different voltages and currents. Such method are usually restricted to fairly simple circuits and include Kirchhoff's law, loop analysis, nodal analysis, superposition theorem, compensation theorem, and reciprocity theorem, etc.

Network Reduction Method

In this method, the original network is converted into a much simpler equivalent circuit for a rapid calculation of different quantities. This method can be applied to a simple as well as complicated network. Examples of this method are: Delta/Star and Star/Delta conversion, Thevenin's theorem, and Norton's theorem, etc.

Parameters of Electrical Circuit

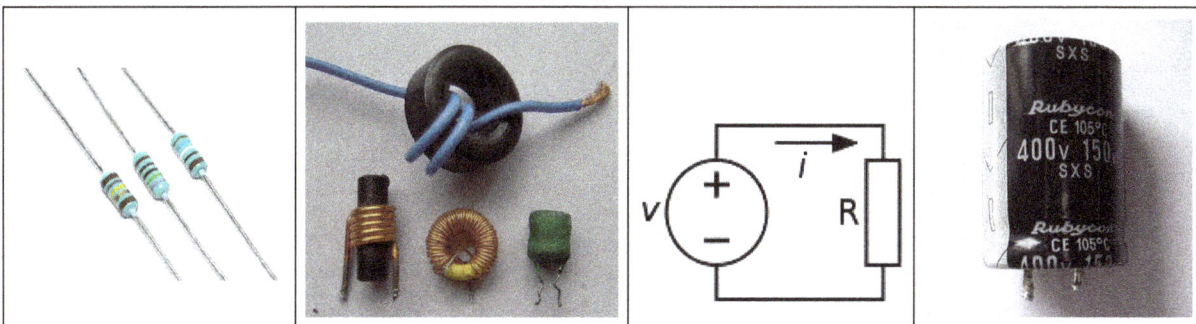

Series and Parallel Electrical Circuits

Resistance Circuits

- Resistance in Series:

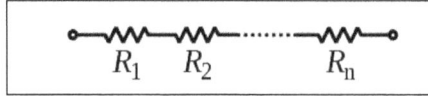

When some resistor of R_1, R_2,....R_n are joined end-on-end as shown, they are said to be connected in series. It can be proved that the equivalent resistance or total resistance between them is equal to the sum of the three individual resistances.

In a series circuit remember that Current is the same through all the conductors but Voltage drop across each is different due to its difference resistance. The total resistance R is equal to,

$$R = R_1 + R_2 + ...+ R_n.$$

- Resistance in Parallel:

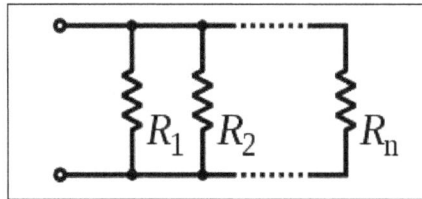

When some resistors of R_1, R_2 ....R_n are joined in parallel as shown in the figure, voltage drop is the same across all the resistors, but the current in each resistor is different and it is given by ohms law. The total current is the sum of the three separate currents so the Equivalent resistance R of all the resistors in Parallel is equal to the reciprocal of the total resistance of the reciprocals of the resistance of their individual resistances.

$$1/R = \{ (1/R_1) + (1/R_2) + ...+ (1/R_n) \}$$

Capacitive Circuits

- Capacitor in Series:

When some of the capacitors, of C_1, C_2.....C_n are connected in series as shown in the figure, then the Equivalent Capacitance C of capacitors in series is equal to the reciprocal of the sum of the reciprocals of their individual capacitances.

$$1/C = \{ (1/C_1) + (1/C_2) + ...+(1/C_n)\}$$

- Capacitor in Parallel:

When some Capacitors, of C_1, C_2 ....Cn are joined in parallel as shown in the figure, then the Equivalent capacitance of all the capacitors connected in parallel is equal to the sum of the individual capacitance.

$$C = C_1 + C_2 + ...+ C_n$$

Inductive Circuits

- Inductors in Series:

When some of the Inductors, of L_1, L_2.....L_n are connected in series as shown, then the Equivalent Inductance L is equal to the sum of their individual inductance connected in the circuit.

$$L = L_1 + L_2 + ... + L_n$$

- Inductors in Parallel:

When some of the Inductors, of L_1, L_2 .....L_n are connected in parallel as shown, then the Equivalent inductance L of inductors in parallel is equal to the reciprocal of the sum of the reciprocals of their individual inductance.

$$1/L = \{ (1/L_1) + (1/L_2) + ... + (1/L_n) \}.$$

CHAPTER 2

Components of Circuits

Some of the major components of circuits are resistors, capacitors, inductors, voltage source, electric current, transistors, thyristors and diodes. They are connected by the conductive wires and traces through which electric current can flow. The topics elaborated in this chapter will help in gaining a better perspective about these components of circuits.

RESISTORS

Special components called resistors are made for the express purpose of creating a precise quantity of resistance for insertion into a circuit. They are typically constructed of metal wire or carbon and engineered to maintain a stable resistance value over a wide range of environmental conditions. Unlike lamps, they do not produce light, but they do produce heat as electric power is dissipated by them in a working circuit. Typically, though, the purpose of a resistor is not to produce usable heat, but simply to provide a precise quantity of electrical resistance.

Resistor Schematic Symbols and Values

The most common schematic symbol for a resistor is a zig-zag line:

Resistor values in ohms are usually shown as an adjacent number, and if several resistors are present in a circuit, they will be labeled with a unique identifier number such as R_1, R_2, R_3, etc. resistor symbols can be shown either horizontally or vertically:

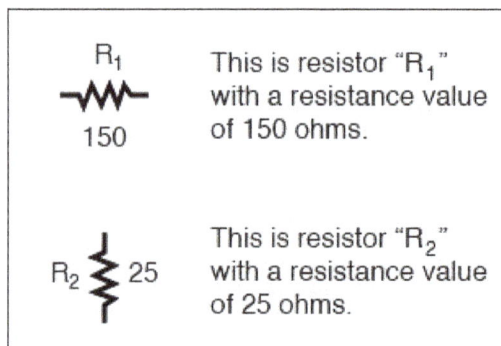

R_1 This is resistor "R_1" with a resistance value of 150 ohms.
150

R_2 25 This is resistor "R_2" with a resistance value of 25 ohms.

Real resistors look nothing like the zig-zag symbol. Instead, they look like small tubes or cylinders with two wires protruding for connection to a circuit. sampling of different kinds and sizes of resistors:

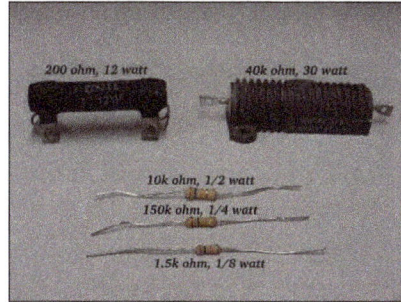

In keeping more with their physical appearance, an alternative schematic symbol for a resistor looks like a small, rectangular box:

Resistors can also be shown to have varying rather than fixed resistances. This might be for the purpose of describing an actual physical device designed for the purpose of providing an adjustable resistance, or it could be to show some component that just happens to have an unstable resistance:

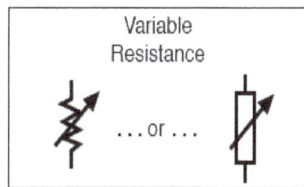

a component symbol drawn with a diagonal arrow through it, that component has a variable rather than a fixed value. This symbol "modifier" (the diagonal arrow) is standard electronic symbol convention.

Variable Resistors

Variable resistors must have some physical means of adjustment, either a rotating shaft or lever that can be moved to vary the amount of electrical resistance. photograph showing some devices called potentiometers, which can be used as variable resistors:

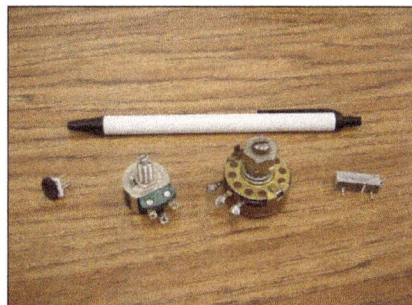

Power Rating of Resistors

Because resistors dissipate heat energy as the electric currents through them overcome the "friction" of their resistance, resistors are also rated in terms of how much heat energy they can dissipate without overheating and sustaining damage. Naturally, this power rating is specified in the physical unit of "watts." Most resistors found in small electronic devices such as portable radios are rated at 1/4 (0.25) watt or less. The power rating of any resistor is roughly proportional to its physical size. Note in the first resistor photograph how the power ratings relate with size: the bigger the resistor, the higher its power dissipation rating. Also, note how resistances (in ohms) have nothing to do with size!

Although it may seem pointless now to have a device doing nothing but resisting electric current, resistors are extremely useful devices in circuits. Because they are simple and so commonly used throughout the world of electricity and electronics.

How are Resistors Useful?

For a practical illustration of resistors' usefulness, examine the photograph below. It is a picture of a printed circuit board, or PCB: an assembly made of sandwiched layers of insulating phenolic fiber-board and conductive copper strips, into which components may be inserted and secured by a low-temperature welding process called "soldering." The various components on this circuit board are identified by printed labels. Resistors are denoted by any label beginning with the letter "R".

This particular circuit board is a computer accessory called a "modem," which allows digital information transfer over telephone lines. There are at least a dozen resistors (all rated at 1/4 watt power dissipation) that can be seen on this modem's board. Every one of the black rectangles (called "integrated circuits" or "chips") contain their own array of resistors for their internal functions, as well.Another circuit board example shows resistors packaged in even smaller units, called "surface mount devices." This particular circuit board is the underside of a personal computer hard disk drive, and once again the resistors soldered onto it are designated with labels beginning with the letter "R".

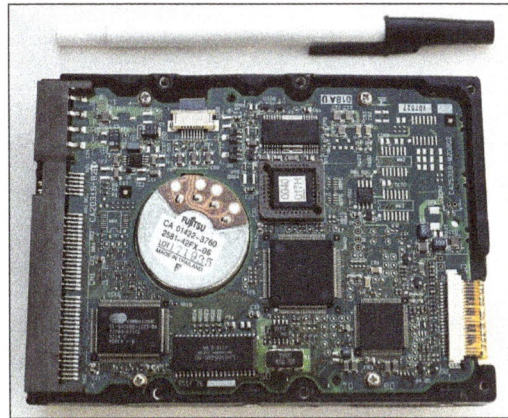

There are over one hundred surface-mount resistors on this circuit board, and this count, of course does not include the number of resistors internal to the black "chips." These two photographs should convince anyone that resistors—devices that "merely" oppose the flow of electric current—are very important components in the realm of electronics.

Load on Schematic Diagrams

In schematic diagrams, resistor symbols are sometimes used to illustrate any general type of device in a circuit doing something useful with electrical energy. Any non-specific electrical device is generally called a load a schematic diagram showing a resistor symbol labeled "load," circuit diagram explaining some concept unrelated to the actual use of electrical power, that symbol may just be a kind of shorthand representation of something else more practical than a resistor.

Analyzing Resistor Circuits

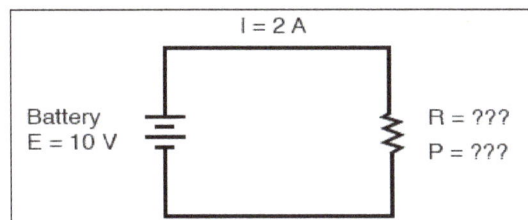

All we've been given here to start with is the battery voltage (10 volts) and the circuit current (2 amps). We don't know the resistor's resistance in ohms or the power dissipated by it in watts. Surveying our array of Ohm's Law equations, we find two equations that give us answers from known quantities of voltage and current:

$$R = \frac{E}{I} \quad \text{and} \quad P = IE$$

Inserting the known quantities of voltage (E) and current (I) into these two equations, determine circuit resistance (R) and power dissipation (P):

$$R \frac{10\,V}{2\,A} = 5\,\Omega$$

$$P=(2A)(10V)=20W$$

For the circuit conditions of 10 volts and 2 amps, the resistor's resistance must be 5 Ω. If we were designing a circuit to operate at these values, we would have to specify a resistor with a minimum power rating of 20 watts, or else it would overheat and fail.

Resistor Materials

Resistors can be found in a variety of different materials, each one with its own properties and specific areas of use. Most electrical engineers use the types found below:

Wirewound (WW) Resistors

Wire Wound Resistors are manufactured by winding resistance wire around a non-conductive core in a spiral. They are typically produced for high precision and power applications. The core is usually made of ceramic or fiberglass and the resistance wire is made of nickel-chromium alloy and are not suitable for applications with frequencies higher than 50kHz. Low noise and stability with respect to temperature variations are standard characteristics of Wire Wound Resistors. Resistance values are available from 0.1 up to 100 kW, with accuracies between 0.1% and 20%.

Metal Film Resistors

Nichrome or tantalum nitride is typically used for metal film resistors. A combination of a ceramic material and a metal typically make up the resistive material. The resistance value is changed by cutting a spiral pattern in the film, much like carbon film with a laser or abrasive. Metal film resistors are usually less stable over temperature than wire wound resistors but handle higher frequencies better.

Metal Oxide Film Resistors

Metal oxide resistors use metal oxides such as tin oxide, making them slightly different from metal film resistors. These resistors are reliable and stable and operate at higher temperatures than metal film resistors. Because of this, metal oxide film resistors are used in applications that require high endurance.

Foil Resistors

Developed in the 1960s, the foil resistor is still one of the most accurate and stable types of resistor that are used for applications with high precision requirements. A ceramic substrate that has a thin bulk metal foil cemented to it makes up the resistive element. Foil Resistors feature a very low-temperature coefficient of resistance.

Carbon Composition (CCR) Resistors

Until the 1960s Carbon Composition Resistors were the standard for most applications. They are reliable, but not very accurate (their tolerance cannot be better than about 5%). A mixture of fine carbon particles and non-conductive ceramic material are used for the resistive element of CCR Resistors. The substance is molded into the shape of a cylinder and baked. The dimensions of the body and the ratio of carbon to ceramic material determine the resistance value. More carbon used in the process means there will be a lower resistance. CCR resistors are still useful for certain applications because of their ability to withstand high energy pulses, a good example application would be in a power supply.

Carbon Film Resistors

Carbon film resistors have a thin carbon film (with a spiral cut in the film to increase the resistive path) on an insulating cylindrical core. This allows for the resistance value to be more accurate and also increases the resistance value. Carbon film resistors are much more accurate than carbon composition resistors. Special carbon film resistors are used in applications that require high pulse stability.

Key Performance Indicators (KPIs)

The KPIs for each resistor material can be found below:

Characteristic	Metal Film	Thick Metal Film	Precision Metal Film	Carbon Composition	Carbon Film
Temp. range	-55+125	-55+130	-55+155	-40+105	.55+155
Max. temp. coeff.	100	100	15	1200	250-1000
Vmax	200-350	250	200	350-500	350-500
Noise (μV per volt of applied DC)	0.5	0.1	0.1	4 (100K)	5 (100K)
R Insul.	10000	10000	10000	10000	10000
Solder (change % in resistance value)	0.20%	0.15%	0.02%	2%	0.50%
Damp heat (change % in resistance value)	0.50%	1%	0.50%	15%	3.50%
Shelf life (change % in resistance value)	0.10%	0.10%	0.00%	5%	2%
Full rating (2000h at 70degC)	1%	1%	0.03%	10%	4%

CAPACITOR

Capacitor is an electronic component that stores electric charge. The capacitor is made of 2 close conductors (usually plates) that are separated by a dielectric material. The plates accumulate electric charge when connected to power source. One plate accumulates positive charge and the other plate accumulates negative charge.

The capacitance is the amount of electric charge that is stored in the capacitor at voltage of 1 Volt.

The capacitance is measured in units of Farad (F).

The capacitor disconnects current in direct current (DC) circuits and short circuit in alternating current (AC) circuits.

Capacitor Symbols

Capacitor	⊶⊣⊢⊶	⊶⊣⊢⊶
Polarized capacitor	⊶⊣⊢⊶	⊶⊣⊢⊶
Variable capacitor	⊶⊣⊢⊶	

Capacitance

The capacitance (C) of the capacitor is equal to the electric charge (Q) divided by the voltage (V):

$$C = \frac{Q}{V}$$

- C is the capacitance in farad (F).

- Q is the electric charge in coulombs (C), that is stored on the capacitor.

- V is the voltage between the capacitor's plates in volts (V).

Capacitance of Plates Capacitor

The capacitance (C) of the plates capacitor is equal to the permittivity (ε) times the plate area (A) divided by the gap or distance between the plates (d):

$$C = \varepsilon \times \frac{A}{d}$$

- C is the capacitance of the capacitor, in farad (F).

- ε is the permittivity of the capacitor's dialectic material, in farad per meter (F/m).

- A is the area of the capacitor's plate in square meters (m²).

- d is the distance between the capacitor's plates, in meters (m).

Capacitors in Series

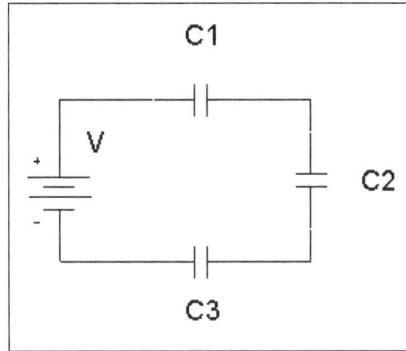

The total capacitance of capacitors in series, $C_1, C_2, C_3, ...$:

$$\frac{1}{C_{Total}} = \frac{1}{C_1} + \frac{1}{C_2} + \frac{1}{C_3} + ...$$

Capacitors in Parallel

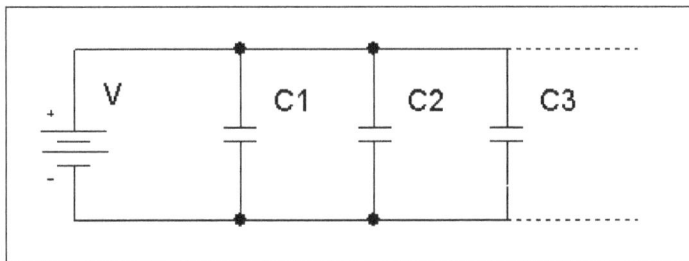

The total capacitance of capacitors in parallel, $C_1, C_2, C_3, ...$:

$$C_{Total} = C_1 + C_2 + C_3 ...$$

Capacitor's Current

The capacitor's momentary current $i_c(t)$ is equal to the capacitance of the capacitor, times the derivative of the momentary capacitor's voltage $v_c(t)$:

$$i_c(t) = C \frac{dv_c(t)}{dt}$$

Capacitor's Voltage

The capacitor's momentary voltage $v_c(t)$ is equal to the initial voltage of the capacitor, plus 1/C times the integral of the momentary capacitor's current $i_c(t)$ over time t:

$$v_c(t) = v_c + \frac{1}{C} \int_0^t i_c(\tau) d\tau$$

Energy of Capacitor

The capacitor's stored energy E_C in joules (J) is equal to the capacitance C in farad (F) times the square capacitor's voltage V_C in volts (V) divided by 2:

$$E_C = C \times V_C^2 / 2$$

AC Circuits

Angular Frequency

$$\omega = 2\pi f$$

- ω - angular velocity measured in radians per second (rad/s).
- f - frequency measured in hertz (Hz).

Capacitor's Reactance

$$X_C = -\frac{1}{\omega_C}$$

Capacitor's Impedance

Cartesian form:

$$Z_C = jX_C = -j\frac{1}{\omega_C}$$

Polar form:

$$Z_C = X_C \angle -90°$$

Capacitor Types

Variable capacitor	Variable capacitor has changeable capacitance.
Electrolytic capacitor	Electrolytic capacitors are used when high capacitance is needed. Most of the electrolytic capacitors are polarized.
Spherical capacitor	Spherical capacitor has a sphere shape.
Power capacitor	Power capacitors are used in high voltage power systems.
Ceramic capacitor	Ceramic capacitor has ceramic dielectric material. Has high voltage functionality.
Tantalum capacitor	Tantalum oxide dielectric material. Has high capacitance.
Mica capacitor	High accuracy capacitors.
Paper capacitor	Paper dielectric material.

INDUCTOR

The inductor is a passive component which stores the electrical energy in the magnetic field when the electric current passes through it. Or say that the inductor is an electrical device which possesses the inductance.

The inductor is made of wire which has the property of inductance, i.e., opposes the flow of current. The inductance of wire increases by increasing the number of turns. The alphabet 'L' is used for representing the inductor, and it is measured in Henry. The inductance characterises the inductor. The figure below shows the symbolic representation of inductor.

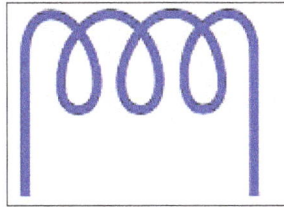

The electric current I flows through the coil generates the magnetic field around it. Consider the magnetic field generates the flux Φ when current flows through it. The ratio of the flux and the current gives inductances.

$$L = \frac{\phi}{I}$$

The inductance of the circuit depends on the current paths and the magnetic permeability of the nearer material. The magnetic permeability shows the ability of the material to forms the magnetic field.

Types of Inductor

The inductors are classified into two types:

1. Air Cored Inductor (wound on non-ferrite material): The inductor in which either the core is completely absent or ceramic material is used for making the core such type of inductor is known as the air-cored inductor.

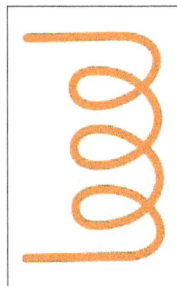

The ceramic material has the very low thermal coefficient of expansion. The low thermal coefficient of expansion means the shape of material remains same even with the rise of temperature.

The ceramic material has no magnetic properties. The permeability of the inductor remains same due to the ceramic material.

In air core-inductor, the only work of the core is to give the coil a particular shape. The air cored structure has many advantages like they reduce the core losses and increases the quality factor. The air cored inductor is used for high-frequency applications work where low inductance is required.

2. Iron Core Inductor (wound on ferrite core): It is a fixed value inductor in which the iron core is kept between the coil. The iron-cored inductor is used in the filter circuit for smoothing out the ripple voltage, it is also used as a choke in fluorescent tube light, in industrial power supplies and inverter system etc.

How Inductor Works?

The inductor is an electrical device used for storing the electrical energy in the form of the magnetic field. It is constructed by wounding the wire on the core. The cores are made of ceramic material, iron or by the air. The core may be toroidal or E- shaped.

The coil-carrying the electric current induces the magnetic field around the conductor. The intensity of the magnetic field increases if the core is placed between the coil. The core provides the low reluctance path to the magnetic flux.

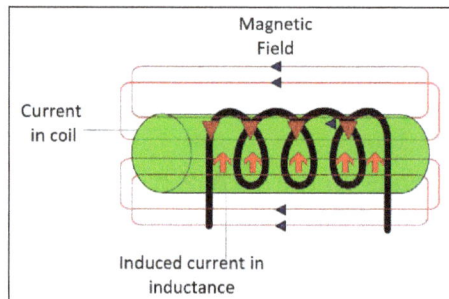

The magnetic field induces the EMF in the coil which causes the current. And according to Lenz's law, the causes always oppose the effect. Here, the current is the causes, and it is induced because of the voltage. Thus, the EMF oppose the change of current that changes the magnetic field. The current which reduces because of the inductance is known as the inductive reactance. The inductive reactance increases with the increase of the number of turn of coils.

VOLTAGE SOURCE

Voltage source is a passive element which can create a continuous force for the movement of electrons through the wire in which it is connected. It is usually a two terminal device.

Types of Voltage Source

- Independent Voltage Source: They are of two types – Direct Voltage Source and Alternating Voltage Source.

- Dependent Voltage Source: They are of two types – Voltage Controlled Voltage Source and Current Controlled Voltage Source.

Independent Voltage Source

The voltage source which can deliver steady voltage (fixed or variable with time) to the circuit and it does not depend on any other elements or quantity in the circuit.

Direct Voltage Source or Time Invariant Voltage Source

The voltage source which can produce or deliver constant voltage as output is termed as Direct Voltage Source. The flow of electrons will be in one direction that is polarity will be always same. The movement of electrons or current will be in one direction always. The value of voltage will not alter with time. Example: DC generator, battery, Cells etc.

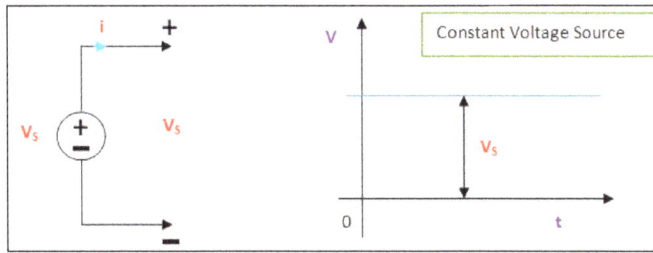

Alternating Voltage Source

The voltage source which can produce or deliver alternating voltage as output is termed as Alternating Voltage Source. Here, the polarity gets reversed at regular intervals. This voltage causes the current to flow in a direction for a time and after that in a different direction for another time. That means it is time varying. Example: DC to AC converter, Alternator etc.

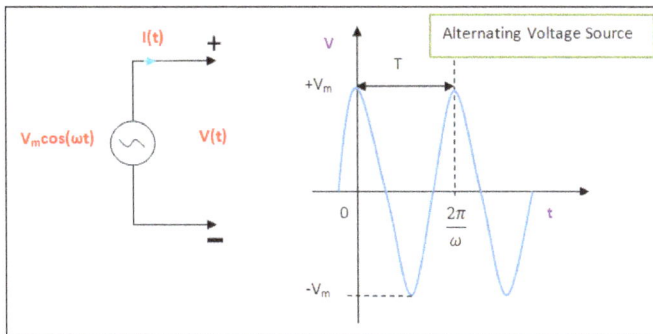

Dependent or Controlled Voltage Source

The voltage source which delivers an output voltage which is not steady or fixed and it always depends on other quantities such as voltage or current in any other part of the circuit is termed as dependent voltage source. They have four terminals. When the voltage source depends on voltage in any other part of the circuit, then it is called Voltage Controlled Voltage Source (VCVS). When the voltage source depends on current in any other part of the circuit, then it is called Current Controlled Voltage Source (CCVS).

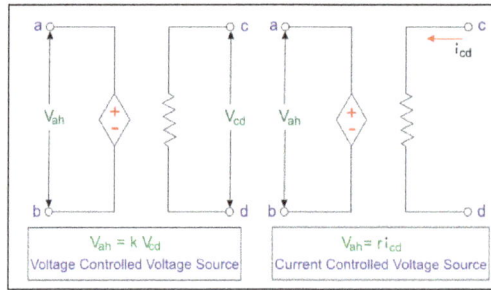

$V_{ah} = k\,V_{cd}$
Voltage Controlled Voltage Source

$V_{ah} = r\,i_{cd}$
Current Controlled Voltage Source

Ideal Voltage Source

The voltage source which can deliver constant voltage to the circuit and it is also referred as independent voltage source as it is independent of the current that the circuit draws. The value of internal resistance is zero here. That is, no power is wasted owing to internal resistance. In spite of the load resistance or current in the circuit, this voltage source will give steady voltage. It performs as a 100% efficient voltage source. All of its voltage of the ideal voltage source can drop perfectly to the load in the circuit.

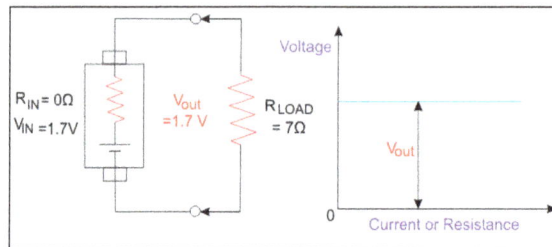

For understanding the ideal voltage source, take an example of a circuit shown above. The battery shown here is an ideal voltage source which delivers 1.7V. The internal resistance R_{IN} = 0Ω. The resistance load in the circuit R_{LOAD} = 7Ω. Here, see the load will receives all of the 1.7V of the battery.

Real or Practical Voltage Source

A circuit with practical voltage source having an internal resistance of 1Ω in the similar circuit which is explained above. Due to the internal resistance, there will be small amount of voltage drop in the R_{IN}. So, the output voltage will be reduced to 1.49V from 1.7V. So in practical cases there will be reduction in source voltage due to the internal resistance the ideal voltage source is kept as models and the real voltage source is made with minimum internal resistance to get the voltage source close to the ideal one with minimum power loss.

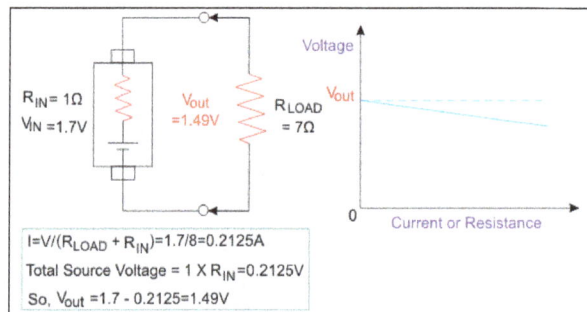

$I = V/(R_{LOAD} + R_{IN}) = 1.7/8 = 0.2125A$

Total Source Voltage $= 1 \times R_{IN} = 0.2125V$

So, $V_{out} = 1.7 - 0.2125 = 1.49V$

ELECTRIC CURRENT

A torch is a simple series circuit with a cell, a switch and a lamp. As soon as you close the switch, the lamp comes on. There is no delay. Let's see why this happens.

An electric current is a flow of electric charge around a circuit. The charge is already in the wires (carried by billions of tiny particles called electrons). This charge is evenly spread out through the wires. As soon as you close the switch, the cell starts to push on the charge. So all the charge starts moving at once.

It's a bit like a bicycle chain. The links are like the charge, the wheel is like the lamp and your feet are like the cell. As soon as you start pedalling, the back wheel starts to move. This is because turning the pedals makes all the links move at once. It's not just the links nearest your feet that move.

Measuring Current

We use an ammeter to measure current. Using two ammeters, we can show that the same current flows all the way round the circuit.

Electric current does not get used up as it flows round the circuit.

The current is measured in amps (A).

Using an Ammeter

An ammeter is connected in series in a circuit. In this way, we can be sure that the current flows through the ammeter.

TRANSISTOR

Transistor is a semiconductor device that can both conduct and insulate. A transistor can act as a switch and an amplifier. It converts audio waves into electronic waves and resistor, controlling electronic current. Transistors have very long life, smaller in size, can operate on lower voltage supplies for greater safety and required no filament current. The first transistor was fabricated with germanium. A transistor performs the same function as a vacuum tube triode, but using semiconductor junctions instead of heated electrodes in a vacuum chamber. It is the fundamental building block of modern electronic devices and found everywhere in modern electronic systems.

Transistor Basics:

A transistor is a three terminal device. Namely,

- Base: This is responsible for activating the transistor.

- Collector: This is the positive lead.

- Emitter: This is the negative lead.

The basic idea behind a transistor is that it lets you control the flow of current through one channel by varying the intensity of a much smaller current that's flowing through a second channel.

Types of Transistors

There are two types of transistors in present; they are bipolar junction transistor (BJT), field effect transistors (FET). A small current is flowing between the base and the emitter; base terminal can control a larger current flow between the collector and the emitter terminals. For a field-effect transistor, it also has the three terminals, they are gate, source, and drain, and a voltage at the gate can control a current between source and drain. The simple diagrams of BJT and FET are shown in figure below:

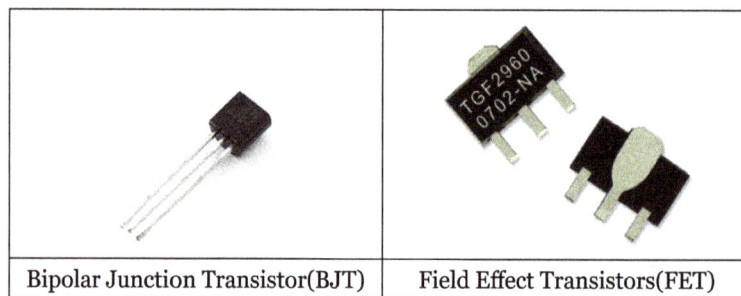

| Bipolar Junction Transistor(BJT) | Field Effect Transistors(FET) |

As you can see, transistors come in a variety of different sizes and shapes. One thing all of these transistors have in common is that they each have three leads.

- Bipolar Junction Transistor

A Bipolar Junction Transistor (BJT) has three terminals connected to three doped semiconductor regions. It comes with two types, P-N-P and N-P-N.

P-N-P transistor, consisting of a layer of N-doped semiconductor between two layers of P-doped material. The base current entering in the collector is amplified at its output.

That is when PNP transistor is ON when its base is pulled low relative to the emitter. The arrows of PNP transistor symbol the direction of current flow when the device is in forward active mode.

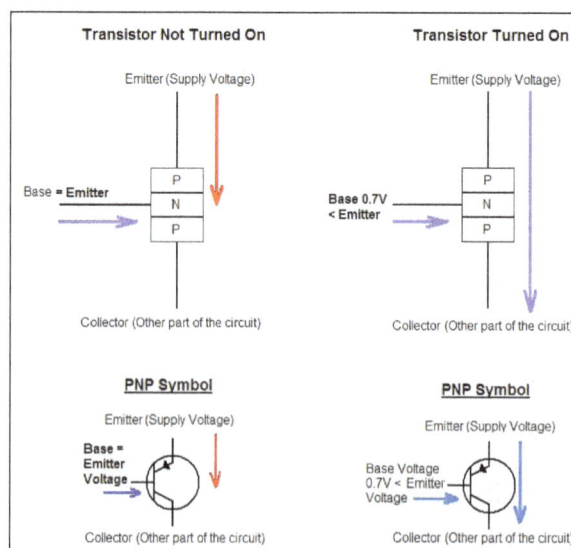

N-P-N transistor consisting a layer of P-doped semiconductor between two layers of N-doped material. By amplifying current the base we get the high collector and emitter current.

That is when NPN transistor is ON when its base is pulled low relative to the emitter. When the transistor is in ON state, current flow is in between the collector and emitter of the transistor. Based on minority carriers in P-type region the electrons moving from emitter to collector. It allows the greater current and faster operation; because of this reason most bipolar transistors used today are NPN.

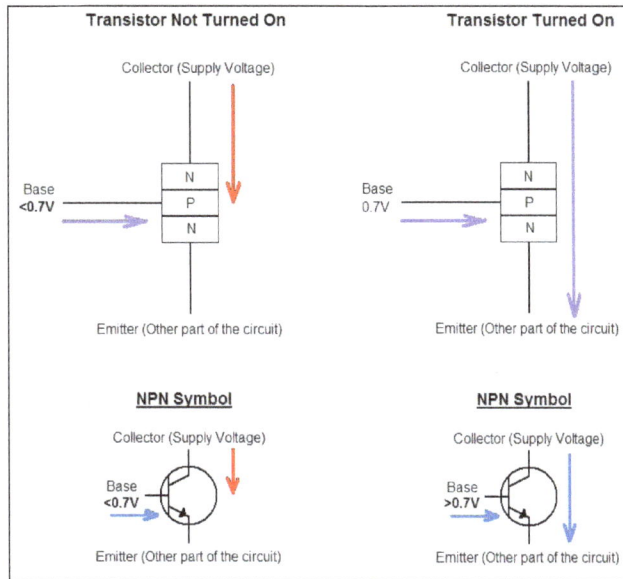

- Field Effect Transistor (FET)

The field-effect transistor is a unipolar transistor, N-channel FET or P-channel FET are used for conduction. The three terminals of FET are source, gate and drain. The basic n-channel and p-channel FET's are shown above. For an n-channel FET, the device is constructed from n-type material. Between the source and drain then-type material acts as a resistor.

This transistor controls the positive and negative carriers with respect to holes or electrons. FET channel is formed by moving of positive and negative charge carriers. The channel of FET which is made by silicon.

There are many types of FET's, MOSFET, JFET and etc. The applications of FET's are in low noise amplifier, buffer amplifier and analog switch.

Bipolar Junction Transistor Biasing

Transistors are the most important semiconductor active devices essential for almost all circuits. They are used as electronic switches, amplifiers etc in circuits. Transistors may be NPN, PNP, FET, JFET etc which have different functions in electronic circuits. For the proper working of the circuit, it is necessary to bias the transistor using resistor networks. Operating point is the point on the output characteristics that shows the Collector-Emitter voltage and the Collector current with no input signal. The Operating point is also known as the Bias point or Q-Point (Quiescent point).

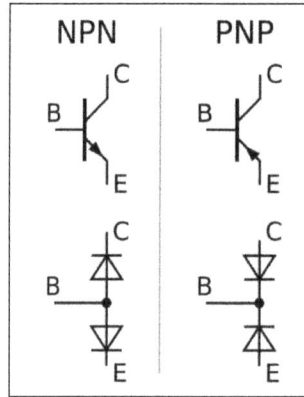

Biasing is referred to provide resistors, capacitors or supply voltage etc to provide proper operating characteristics of the transistors. DC biasing is used to obtain DC collector current at a particular collector voltage. The value of this voltage and current are expressed in terms of the Q-Point. In a transistor amplifier configuration, the IC (max) is the maximum current that can flow through the transistor and VCE (max) is the maximum voltage applied across the device. To work the transistor as an amplifier, a load resistor RC must be connected to the collector. Biasing set the DC operating voltage and current to the correct level so that the AC input signal can be properly amplified by the transistor. The correct biasing point is somewhere between the fully ON or fully OFF states of the transistor. This central point is the Q-Point and if the transistor is properly biased, the Q-point will be the central operating point of the transistor. This helps the output current to increase and decrease as the input signal swings through the complete cycle.

For setting the correct Q-Point of the transistor, a collector resistor is used to set the collector current to a constant and steady value without any signal in its base. This steady DC operating point is set by the value of the supply voltage and the value of the base biasing resistor. Base bias resistors are used in all the three transistor configurations like common base, common collector and Common emitter configurations.

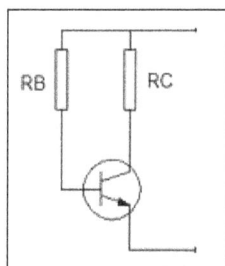

Current Biasing

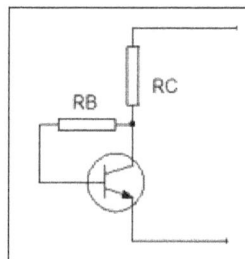

Feedback Biasing

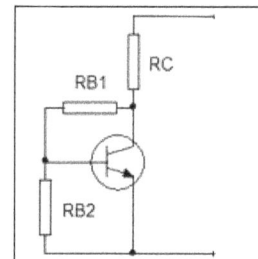

Double Feedback Biasing

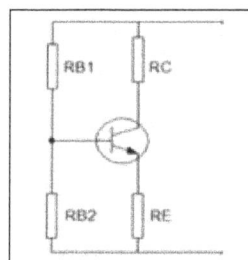

Voltage Divider Biasing

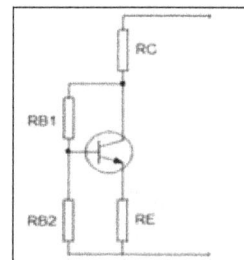

Double Base Biasing

Modes of Biasing

Following are the different modes of transistor base biasing:

Current Biasing

In the figure, two resistors RC and RB are used to set the base bias. These resistors establish the initial operating region of the transistor with a fixed current bias.

The transistor forward biases with a positive base bias voltage through RB. The forward base-Emitter voltage drop is 0.7 volts. Therefore the current through RB is $I_B = (V_{cc} - V_{BE}) / I_B$.

Feedback Biasing

Figure shows the transistor biasing by the use of a feedback resistor. The base bias is obtained from the collector voltage. The collector feedback ensures that the transistor is always biased in the active region. When the collector current increases, the voltage at the collector drops. This reduces the base drive which in turn reduces the collector current. This feedback configuration is ideal for transistor amplifier designs.

Double Feedback Biasing

Figure shows how the biasing is achieved using double feedback resistors. By using two resistors RB1 and RB2 increases the stability with respect to the variations in Beta by increasing the current flow through the base bias resistors. In this configuration, the current in RB1 is equal to 10 % of the collector current.

Voltage Dividing Biasing

Figure shows the Voltage divider biasing in which two resistors RB1 and RB2 are connected to the base of the transistor forming a voltage divider network. The transistor gets biases by the voltage drop across RB2. This kind of biasing configuration is used widely in amplifier circuits.

Double Base Biasing

Figure shows a double feedback for stabilization. It uses both Emitter and Collector base feedback to improve the stabilization through controlling the collector current. Resistor values should be selected so as to set the voltage drop across the Emitter resistor 10% of the supply voltage and the current through RB1, 10% of the collector current.

Advantages of Transistor

1. Smaller mechanical sensitivity.

2. Lower cost and smaller in size, especially in small-signal circuits.

3. Low operating voltages for greater safety, lower costs and tighter clearances.

4. Extremely long life.

5. No power consumption by a cathode heater.

6. Fast switching.

THYRISTOR

It is a multi-layer semiconductor device, hence the "silicon" part of its name. It requires a gate signal to turn it "ON", the "controlled" part of the name and once "ON" it behaves like a rectifying diode, the "rectifier" part of the name. In fact the circuit symbol for the thyristor suggests that this device acts like a controlled rectifying diode.

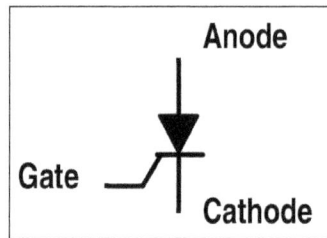

Thyristor Symbol.

However, unlike the junction diode which is a two layer (P-N) semiconductor device, or the commonly used bipolar transistor which is a three layer (P-N-P, or N-P-N) switching device, the Thyristor is a four layer (P-N-P-N) semiconductor device that contains three PN junctions in series, and is represented by the symbol.

Like the diode, the Thyristor is a unidirectional device, that is it will only conduct current in one direction only, but unlike a diode, the thyristor can be made to operate as either an open-circuit switch or as a rectifying diode depending upon how the thyristors gate is triggered. In other words, thyristors can operate only in the switching mode and cannot be used for amplification.

The silicon controlled rectifier SCR, is one of several power semiconductor devices along with Triacs (Triode AC's), Diacs (Diode AC's) and UJT's (Unijunction Transistor) that are all capable of acting like very fast solid state AC switches for controlling large AC voltages and currents. So for the Electronics student this makes these very handy solid state devices for controlling AC motors, lamps and for phase control.

The thyristor is a three-terminal device labelled: "Anode", "Cathode" and "Gate" and consisting of three PN junctions which can be switched "ON" and "OFF" at an extremely fast rate, or it can be switched "ON" for variable lengths of time during half cycles to deliver a selected amount of power to a load. The operation of the thyristor can be best explained by assuming it to be made up of two transistors connected back-to-back as a pair of complementary regenerative switches.

A Thyristors two Transistor Analogy

The two transistor equivalent circuit shows that the collector current of the NPN transistor TR_2 feeds directly into the base of the PNP transistor TR_1, while the collector current of TR_1 feeds into the base of TR_2. These two inter-connected transistors rely upon each other for conduction as each transistor gets its base-emitter current from the other's collector-emitter current. So until one of the transistors is given some base current nothing can happen even if an Anode-to-Cathode voltage is present.

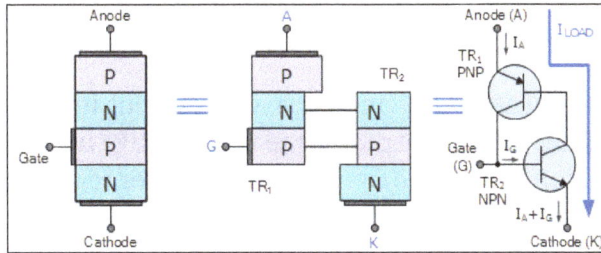

When the thyristors Anode terminal is negative with respect to the Cathode, the centre *N-P*junction is forward biased, but the two outer *P-N* junctions are reversed biased and it behaves very much like an ordinary diode. Therefore a thyristor blocks the flow of reverse current until at some high voltage level the breakdown voltage point of the two outer junctions is exceeded and the thyristor conducts without the application of a Gate signal.

This is an important negative characteristic of the thyristor, as Thyristors can be unintentionally triggered into conduction by a reverse over-voltage as well as high temperature or a rapidly rising *dv/dt* voltage such as a spike.

If the Anode terminal is made positive with respect to the Cathode, the two outer *P-N*junctions are now forward biased but the centre *N-P* junction is reverse biased. Therefore forward current is also blocked. If a positive current is injected into the base of the NPN transistor TR$_2$, the resulting collector current flows in the base of transistor TR$_1$. This in turn causes a collector current to flow in the PNP transistor, TR$_1$ which increases the base current of TR$_2$ and so on.

Very rapidly the two transistors force each other to conduct to saturation as they are connected in a regenerative feedback loop that can not stop. Once triggered into conduction, the current flowing through the device between the Anode and the Cathode is limited only by the resistance of the external circuit as the forward resistance of the device when conducting can be very low at less than 1Ω so the voltage drop across it and power loss is also low

Typical Thyristor.

A thyristor blocks current in both directions of an AC supply in its "OFF" state and can be turned "ON" and made to act like a normal rectifying diode by the application of a positive current to the base of transistor, TR$_2$ which for a silicon controlled rectifier is called the "Gate" terminal.

The operating voltage-current I-V characteristics curves for the operation of a Silicon Controlled Rectifier are given as:

Thyristor I-V Characteristics Curves

Once the thyristor has been turned "ON" and is passing current in the forward direction (anode positive), the gate signal looses all control due to the regenerative latching action of the two internal transistors. The application of any gate signals or pulses after regeneration is initiated will have no effect at all because the thyristor is already conducting and fully-ON.

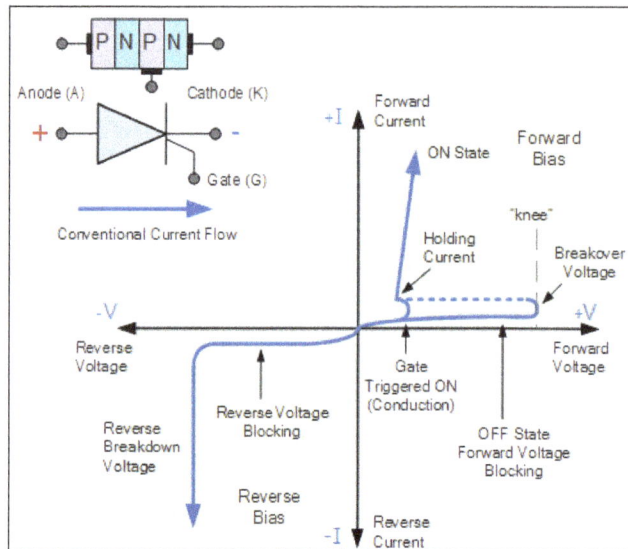

Unlike the transistor, the SCR can not be biased to stay within some active region along a load line between its blocking and saturation states. The magnitude and duration of the gate "turn-on" pulse has little effect on the operation of the device since conduction is controlled internally. Then applying a momentary gate pulse to the device is enough to cause it to conduct and will remain permanently "ON" even if the gate signal is completely removed.

Therefore the thyristor can also be thought of as a *Bistable Latch* having two stable states "OFF" or "ON". This is because with no gate signal applied, a silicon controlled rectifier blocks current in both directions of an AC waveform, and once it is triggered into conduction, the regenerative latching action means that it cannot be turned "OFF" again just by using its Gate.

So how do we turn "OFF" the thyristor?. Once the thyristor has self-latched into its "ON" state and passing a current, it can only be turned "OFF" again by either removing the supply voltage and therefore the Anode (I_A) current completely, or by reducing its Anode to Cathode current by some external means (the opening of a switch for example) to below a value commonly called the "minimum holding current", I_H.

The Anode current must therefore be reduced below this minimum holding level long enough for the thyristors internally latched pn-junctions to recover their blocking state before a forward voltage is again applied to the device without it automatically self-conducting. Obviously then for a thyristor to conduct in the first place, its Anode current, which is also its load current, I_L must be greater than its holding current value. That is $I_L > I_H$.

Since the thyristor has the ability to turn "OFF" whenever the Anode current is reduced below this minimum holding value, it follows then that when used on a sinusoidal AC supply the SCR will automatically turn itself "OFF" at some value near to the cross over point of each half cycle, and as will remain "OFF" until the application of the next Gate trigger pulse.

Since an AC sinusoidal voltage continually reverses in polarity from positive to negative on every half-cycle, this allows the thyristor to turn "OFF" at the 180° zero point of the positive waveform. This effect is known as "natural commutation" and is a very important characteristic of the silicon controlled rectifier.

Thyristors used in circuits fed from DC supplies, this natural commutation condition cannot occur as the DC supply voltage is continuous so some other way to turn "OFF" the thyristor must be provided at the appropriate time because once triggered it will remain conducting.

However in AC sinusoidal circuits natural commutation occurs every half cycle. Then during the positive half cycle of an AC sinusoidal waveform, the thyristor is forward biased (anode positive) and a can be triggered "ON" using a Gate signal or pulse. During the negative half cycle, the Anode becomes negative while the Cathode is positive. The thyristor is reverse biased by this voltage and cannot conduct even if a Gate signal is present.

So by applying a Gate signal at the appropriate time during the positive half of an AC waveform, the thyristor can be triggered into conduction until the end of the positive half cycle. Thus phase control (as it is called) can be used to trigger the thyristor at any point along the positive half of the AC waveform and one of the many uses of a Silicon Controlled Rectifier is in the power control of AC systems.

Thyristor Phase Control

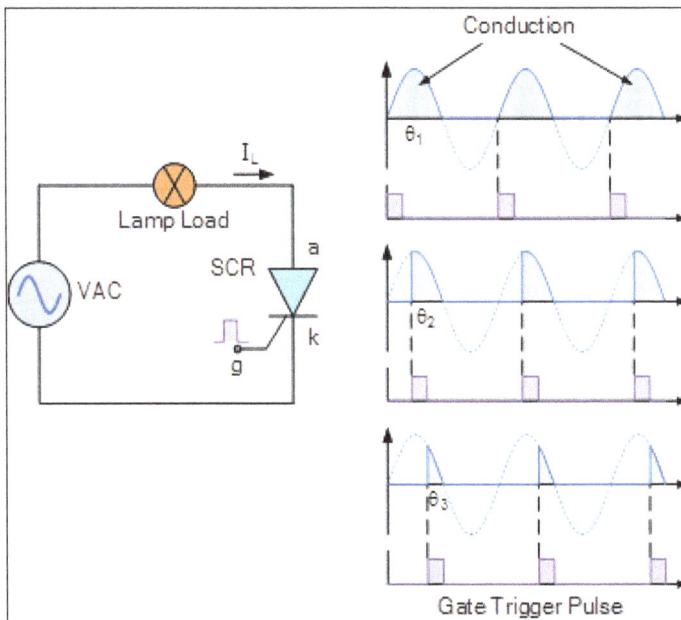

At the start of each positive half-cycle the SCR is "OFF". On the application of the gate pulse triggers the SCR into conduction and remains fully latched "ON" for the duration of the positive cycle.

If the thyristor is triggered at the beginning of the half-cycle ($\theta = 0°$), the load (a lamp) will be "ON" for the full positive cycle of the AC waveform (half-wave rectified AC) at a high average voltage of 0.318 x Vp.

As the application of the gate trigger pulse increases along the half cycle ($\theta = 0°$ to $90°$), the lamp is illuminated for less time and the average voltage delivered to the lamp will also be proportionally less reducing its brightness.

Then we can use a silicon controlled rectifier as an AC light dimmer as well as in a variety of other AC power applications such as: AC motor-speed control, temperature control systems and power regulator circuits, etc.

This far we have seen that a thyristor is essentially a half-wave device that conducts in only the positive half of the cycle when the Anode is positive and blocks current flow like a diode when the Anode is negative, irrespective of the Gate signal.

But there are more semiconductor devices available which come under the banner of "Thyristor" that can conduct in both directions, full-wave devices, or can be turned "OFF" by the Gate signal.

Such devices include "Gate Turn-OFF Thyristors" (GTO), "Static Induction Thyristors" (SITH), "MOS Controlled Thyristors" (MCT), "Silicon Controlled Switch" (SCS), "Triode Thyristors" (TRI-AC) and "Light Activated Thyristors" (LASCR) to name a few, with all these devices available in a variety of voltage and current ratings making them attractive for use in applications at very high power levels.

Silicon Controlled Rectifiers known commonly as Thyristors are three-junction PNPN semiconductor devices which can be regarded as two inter-connected transistors that can be used in the switching of heavy electrical loads. They can be latched-"ON" by a single pulse of positive current applied to their Gate terminal and will remain "ON" indefinitely until the Anode to Cathode current falls below their minimum latching level.

Static Characteristics of a Thyristor

- Thyristors are semiconductor devices that can operate only in the switching mode.

- Thyristor are current operated devices, a small Gate current controls a larger Anode current.

- Conducts current only when forward biased and triggering current applied to the Gate.

- The thyristor acts like a rectifying diode once it is triggered "ON".

- Anode current must be greater than holding current to maintain conduction.

- Blocks current flow when reverse biased, no matter if Gate current is applied.

- Once triggered "ON", will be latched "ON" conducting even when a gate current is no longer applied providing Anode current is above latching current.

Thyristors are high speed switches that can be used to replace electromechanical relays in many circuits as they have no moving parts, no contact arcing or suffer from corrosion or dirt. But in addition to simply switching large currents "ON" and "OFF", thyristors can be made to control the mean value of an AC load current without dissipating large amounts of power. A good example of thyristor power control is in the control of electric lighting, heaters and motor speed.

DIODE

The diode is the first semiconductor device. The diode's distinctive feature is that it conducts current in one direction, but not the other. We won't go into the details of how a diode does this, or how it's made. Fortunately, you don't have to know how to make a diode before using it in a circuit.

Where we're Headed

The $i - v$ curve of a diode is modeled by this non-linear equation:

$$i = I_S(e^{\,qv/kT} - 1)$$

I_S is the reverse saturation current. For silicon, typically $10^{-12}\,A$.

q is the charge on an electron, $1.602 \times 10^{-19}\,C$.

k is Boltzmann's constant, $1.380 \times 10^{-23}\,J/K$

T is the temperature in kelvin.

- We will define terms like forward bias, reverse bias, and saturation current.

- You will learn some tips for identifying the terminals of a real-world diode.

- We will solve a diode circuit using a graphical method.

Diode Symbol

The schematic symbol for a diode looks like this:

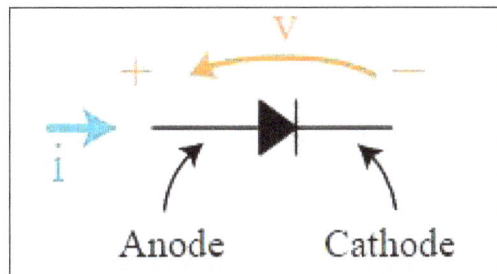

The black arrow ▶ in the symbol points in the direction of the diode's forward current, i, the direction where current flow happens. The diode's voltage, v, is oriented with the + sign on the end where forward current comes into the diode. We use the sign convention for passive components. The optional curved orange arrow also indicates the voltage polarity.

Diode i - v Curve

This is a typical $i-v$ curve for a silicon diode. A diode is a non-linear device:

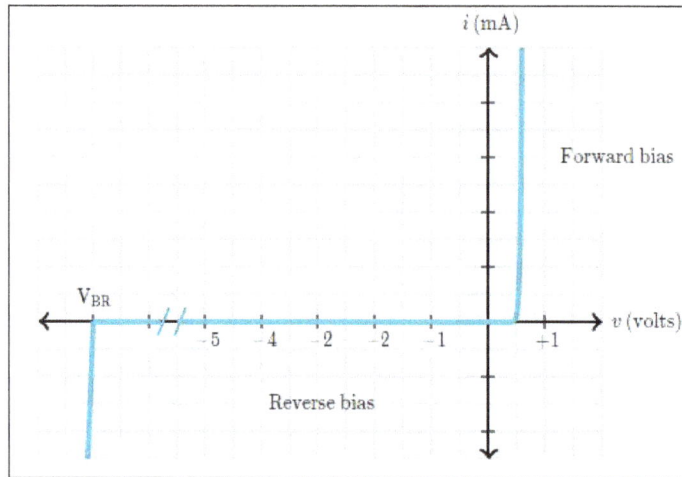

Diode $i-v$ curve of a silicon diode. A positive voltage means the diode is forward biased.
A negative voltage means the diode is operating with reverse bias.

Forward and Reverse Current

Forward Current

Let's say we place a very small positive voltage, like +0.2 olts, across a silicon diode. That puts us on the right side of the $i-v$ curve. With this small positive voltage, almost no forward current flows. When the voltage increases up to around $0.6\,\text{V}$ measurable current starts to flow through the diode in the forward direction. As the voltage moves a little above $0.6\,\text{V}$, the current through the diode rises rapidly. The $i-v$ curve is nearly vertical at this point.

With a positive voltage on its terminals, we say the diode is forward biased. A diode is forward biased when its voltage is anywhere on the + voltage side of the origin. In normal operation, the voltage across a forward biased silicon diode is somewhere between $0.60 - 0.75\text{V}$ If you externally force the voltage higher than 0.75 volts, the diode current gets very large and it may overheat.

Reverse Current

If you put a negative voltage to a diode, so the - terminal is at a higher voltage than the + terminal, this puts us over on the left side of the $i-v$ curve. We say the diode is reverse biased. In the reverse direction, the current is very close to zero, just ever so slightly negative, below the voltage axis.

A reverse biased diode can't hold out forever. When the voltage reaches a high negative value known as the breakdown voltage, V_{BR} the diode starts to conduct in the reverse direction. At breakdown, the current sharply increases and becomes very high in the negative direction. A breakdown voltage V_{BR} of -50V is typical of ordinary diodes. Most of the time you don't allow the diode voltage to get near V_{BR}.

Diode Terminals

When you draw diodes, the schematic symbol clearly indicates the direction of forward current flow.

You don't really need names for the two terminals. But, if you are handling real diodes to build a circuit, you have to figure out which way to point the diode. Real diodes are so small there isn't room to paint a little diode symbol on them, so you need to identify the terminals some other way.

The two terminals of a diode are the anode and cathode.

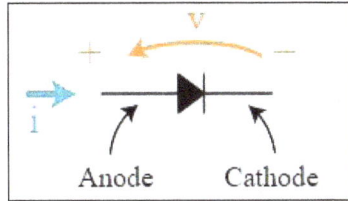

How can you Remember the Anode and Cathode?

For the longest time you could not remember which end of the diode was called the anode and which was the cathode, you would be looked it up every time. you will finally came up with this memory aid. The word for cathode is Kathode. The big K kind of looks like a diode symbol.

Flip the diode symbol around until it reads like a K. The Kathode is the terminal on the left.

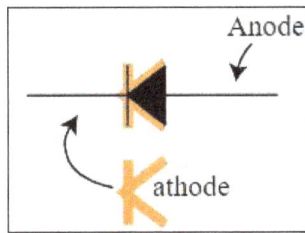

Identifying the Terminals of a Real-life Diode

Diodes are made on small chips of silicon. They are delivered to you in all sorts of tiny packages. There are a few different ways to indicate which diode terminal is which.

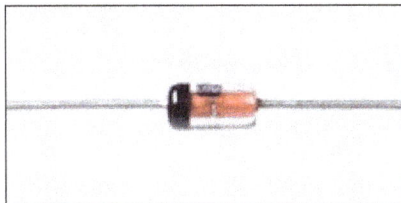

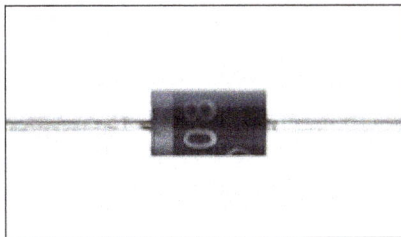

Diode packages like the glass and black plastic cylinders shown above usually have a painted bar near one end. The bar on the package is the bar of the diode symbol, so it indicates the cathode.

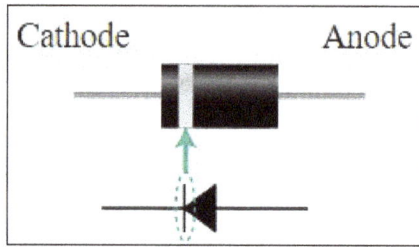

The stripe (any contrasting color) corresponds to the diode's cathode.

This red LED (light emitting diode) has wire leads of different length. The forward current goes into the longer lead (anode). The package may have a bump or tab sticking out on the forward current side.

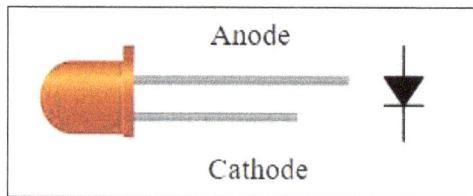

The longer lead corresponds to the diode's anode.

Identify the Anode and Cathode with a Meter

A good way to verify the identity of the terminals is using an ohm-meter to figure out the forward current direction. On the resistance setting, Ω, the meter puts a small voltage on its test leads (this is why an ohm meter needs a battery). You use that small voltage to see which way current flows.

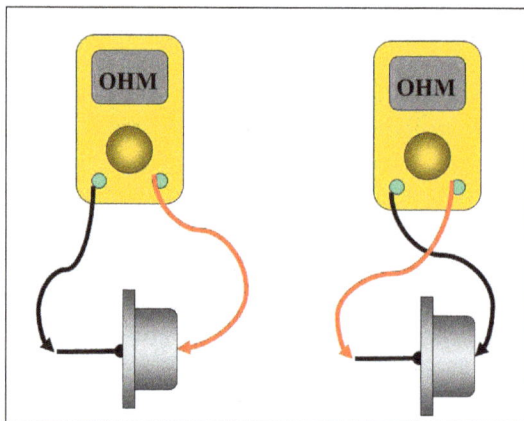

The diode is flipped in each image. If the ohm-meter reads a finite resistance, that means the diode is conducting a small current in the forward direction, and the red + lead from the meter is touching the anode. If the resistance reads O.L (for overload), the diode is not conducting current. That means the red + test lead is touching the cathode.

Your meter might have a diode setting, a little diode symbol. If it does, it will display the forward voltage when the red lead is touching the forward current terminal (the anode) as shown in figure.

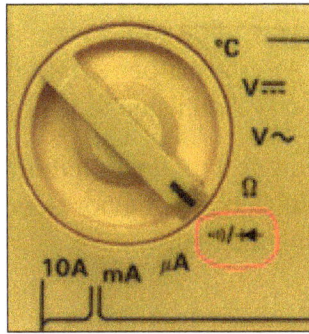

Diode i - v Equation

The diode $i-v$ relationship can be modeled with an equation. This equation is based on the physics underlying the diode action, along with careful measurements on real diodes.

$$i = I_S(e^{\,qv/kT}-1)$$

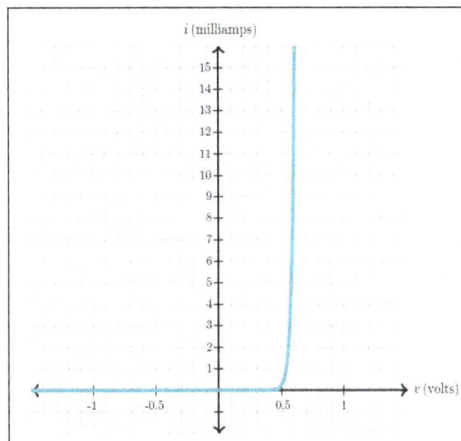

The i-v curve for a typical silicon diode.

The plot above doesn't look very much like an exponential curve, and the current for negative voltages appears to be 0. If we expand the current scale a whole bunch (milliamperes → picoamperes) the exponential shape becomes apparent (the voltage scale is expanded, to). You can see tiny negative I_S flows when the diode is reverse biased:

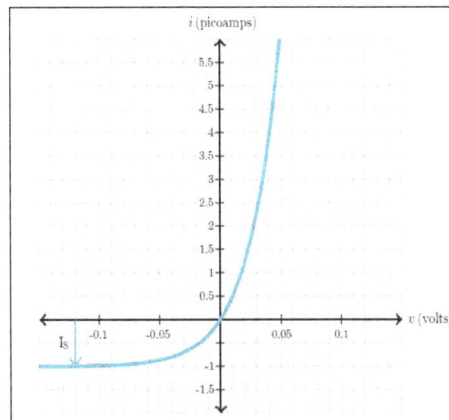

I_S is the saturation current. This current flows backwards when the diode is reverse biased. A typical value for I_S in silicon is 10^{-12} A[10], start superscript, minus, 12, end superscript, space, A, (1 picoampere). For germanium diodes, a typical value for I_S is 10^{-6} A, (1 microampere).

It is best to think of this diode equation as a model of a diode, rather than as a law. The equation represents an abstract ideal diode. The actual behavior depends on how it is made, its temperature, and how much you care about the fine details.

Detailed Look Inside the Diode i - v equation

This next part takes apart the diode equation in some detail. You don't need this to use a diode in a circuit.

There are many new parameters in the diode equation. Let's go through them carefully.

$$i = I_S(e^{qv/kT} - 1)$$

v is the voltage across the diode. We find it up top in the exponential term, which explains why current i has an exponential dependence on voltage v.

Now lets look at all that other stuff up in the exponent of $e^{qv/kT}$.

We know exponents have no dimensions, so the other terms in the exponent have to end up with units of $1/v$.

q is the charge on an electron, in coulombs:

$$q = 1.602 \times 10^{-19} C.$$

k is Boltzmann›s constant, a very important number in physics. The energy of a particle increases with temperature. If you know the temperature of a particle, k tells you how much kinetic energy the particle has just by virtue of being warm. The units of Boltzmann›s constant are energy per kelvin.

$$k = 1.380 \times 10^{-23} J / K \text{(joules per kelvin)}$$

T is the temperature measured from absolute zero in kelvin or K. A temperature of absolute zero, or $0 K$ is $-273° C$(celsius).

If a particle is at $T = 300 K$, (room temperature), then it has energy:

$$kT = 1.380 \times 10^{-23} \text{ J/ K} \cdot 300 K = 4.14 \times 10^{-21} J$$

If the particle is an electron, it has a known charge, and we can talk about its energy per charge. Energy per charge might sound familiar. Its other name is voltage.

$$\frac{kT}{q} = \frac{4.14 \times 10^{-21} J}{1.602 \times 10^{-19} C} = 25.8 \approx 26 mV$$

At room temperature (around 300 K), kT/q is 26 millivolts. That's the energy of a normal everyday electron. The exponent of the diode equation, $v/26\,\text{mV}$, is comparing the diode voltage to the energy of an ordinary electron.

If you feel like it, you can write the diode equation for room temperature as:

$$i = I_S(e^{qv/kT} - 1)$$

This non-linear diode $i-v$ equation is harder to deal with than the linear $i-v$ equations for $R, L, \text{and } C$. There are very few cases where you will be asked to use this equation to find an analytical solution. The usual approach to diode circuits is to perform a graphical solution or to use a circuit simulation program to get an approximate answer.

References

- Resistors, chpt-2, direct-current, textbook: allaboutcircuits.com, Retrieved 2 June, 2019

- Capacitor, electric: rapidtables.com, Retrieved 22 May, 2019

- Inductor: circuitglobe.com, Retrieved 13 March, 2019

- Voltage-source: electrical4u.com, Retrieved 29 June 2019

- Transistors-basics-types-baising-modes: elprocus.com, Retrieved 30 January, 2019

- Thyristor, power: electronics-tutorials.ws, Retrieved 31 March, 2019

- Ee-diode-circuit-element, ee-diode, ee-semiconductor-devices, electrical-engineering, science: khanacademy.org, Retrieved 14 July, 2019

Types of Circuits

Circuits can be classified into various types such as electrical circuits, AC circuits, DC circuits and electronic circuits. DC circuits and AC circuits involve the flow of DC current and AC current respectively. These types of circuits as well as the various subtypes of electronic circuits have been thoroughly discussed in this chapter.

ELECTRICAL CIRCUIT

An electrical circuit is a path or line through which an electrical current flows. The path may be closed (joined at both ends), making it a loop. A closed circuit makes electrical current flow possible. It may also be an open circuit where the electron flow is cut short because the path is broken. An open circuit does not allow electrical current to flow.

Below is a basic set of symbols that you may find on circuit diagrams.

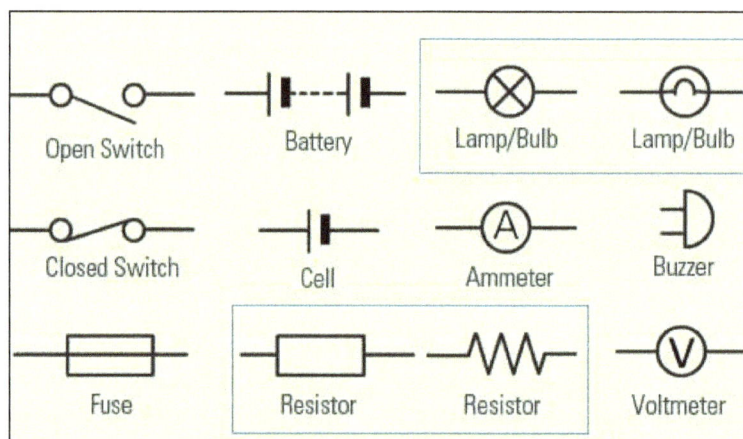

It is very important to know the basic parts of a simple circuit and the symbols that relate to them. A simple circuit has conductors, a switch, a load and a power source. Here are the functions of each part:

- Conductors: These are usually copper wires with no insulation. They make the path through which the electricity flows. One piece of the wire connects the current from the power source (cell) to the load. The other piece connects the load back to the power source.

- Switch: The switch is simply a small gap in the conductor where you can close or open the circuit. When the switch is closed, the circuit is closed and electricity flows.

- The Load: The load is a small light bulb or buzzer that lights when the circuit is turned on. The load is also known as a resistor.

- Cell: The power source is a cell. (Note that more than one cell put together is known as a battery).

The diagram below shows how a basic circuit looks like.

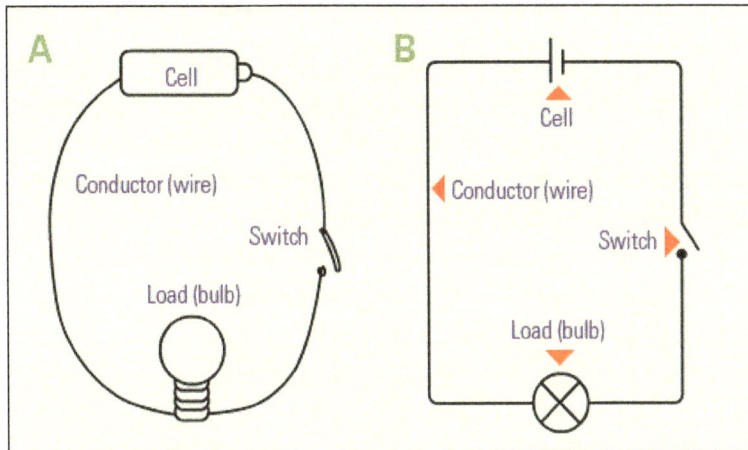

It is important to draw circuits with clean straight lines, as shown in diagram. Avoid realistic sketches. It is important to know that a circuit can have more than the basic components in the diagram. It can have two or more batteries or two or more bulbs.

AC CIRCUIT

The path for the flow of alternating current is called an AC Circuit. The alternating current (AC) is used for domestic and industrial purposes. In an AC circuit, the value of the magnitude and the direction of current and voltages is not constant, it changes at a regular interval of time. It travels as a sinusoidal wave completing one cycle as half positive and half negative cycle and is a function of time (t) or angle (θ=wt).

In DC Circuit, the opposition to the flow of current is the only resistance of the circuit whereas the opposition to the flow of current in the AC circuit is because of resistance (R), Inductive Reactance (X_L=2πfL) and capacitive reactance (X_C = 1/2 πfC) of the circuit.

In AC Circuit, the current and voltages are represented by magnitude and direction. The alternating quantity may or may not be in phase with each other depending upon the various parameters of the circuit like resistance, inductance, and capacitance. The sinusoidal alternating quantities are voltage and current which varies according to the sine of angle θ.

For the generation of electric power, in all over the world the sinusoidal voltage and current are selected because of the following reasons are given below.

- The sinusoidal voltage and current produce low iron and copper losses in the transformer and rotating electrical machines, which in turns improves the efficiency of the AC machines.

- They offer less interference to the nearby communication system.

- They produce less disturbance in the electrical circuit.

Alternating Voltage and Current in an AC Circuit

The voltage that changes its polarity and magnitude at regular interval of time is called an alternating voltage. Similarly the direction of the current is changed and the magnitude of current changes with time it is called alternating current.When an alternating voltage source is connected across a load resistance as shown in the figure below, the current through it flows in one direction and then in the opposite direction when the polarity is reversed.

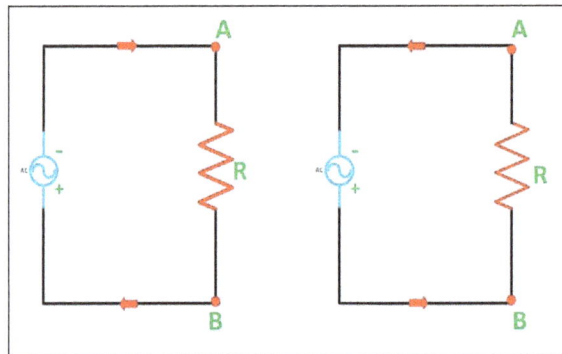

Alternating Currnt Circuit Diagram.

The waveform of the alternating voltage with respect to the time and the current flowing through the resistance (R) in the circuit is shown below.

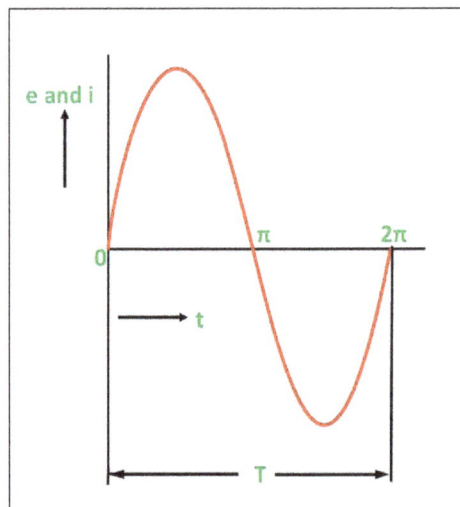

There are various types of AC circuit such as AC circuit containing only resistance (R), AC circuit containing only capacitance (C), AC circuit containing only inductance (L), the combination of RL Circuit, AC circuit containing resistance and capacitance (RC), AC circuit containing inductance and capacitance (LC) and resistance inductance and capacitance (RLC) AC circuit.

The various terms which are frequently used in an AC Circuit are as follows:

Amplitude

The maximum positive or negative value attained by an alternating quantity in one complete cycle is called Amplitude or peak value or maximum value. The maximum value of voltage and current is represented by E_m or V_m and I_m respectively.

Alternation

One half cycle is termed as alternation. An alternation span is of 180 degrees electrical.

Cycle

When one set of positive and negative values completes by an alternating quantity or it goes through 360 degrees electrical, it is said to have one complete Cycle.

Instantaneous Value

The value of voltage or current at any instant of time is called an instantaneous value.It is denoted by (i or e).

Frequency

The number of cycles made per second by an alternating quantity is called frequency. It is measured in cycle per second (c/s) or hertz (Hz) and is denoted by (f).

Time Period

The time taken in seconds by a voltage or a current to complete one cycle is called Time Period. It is denoted by (T).

Wave Form

The shape obtained by plotting the instantaneous values of an alternating quantity such as voltage and current along the y axis and the time (t) or angle ($\theta=wt$) along the x axis is called waveform.

Alternating Current

Alternating current (AC) is an electric current which periodically reverses direction, in contrast to direct current (DC) which flows only in one direction. Alternating current is the form in which electric power is delivered to businesses and residences, and it is the form of electrical energy that consumers typically use when they plug kitchen appliances, televisions, fans and electric lamps into a wall socket. A common source of DC power is a battery cell in a flashlight. The abbreviations *AC* and *DC* are often used to mean simply *alternating* and *direct*, as when they modify *current* or *voltage*.

The usual waveform of alternating current in most electric power circuits is a sine wave, whose positive half-period corresponds with positive direction of the current and vice versa. In certain applications, like guitar amplifiers, different waveforms are used, such as triangular or square waves. Audio and radio signals carried on electrical wires are also examples of alternating current. These types of alternating current carry information such as sound (audio) or images (video) sometimes carried by modulation of an AC carrier signal. These currents typically alternate at higher frequencies than those used in power transmission.

Transmission, Distribution and Domestic Power Supply

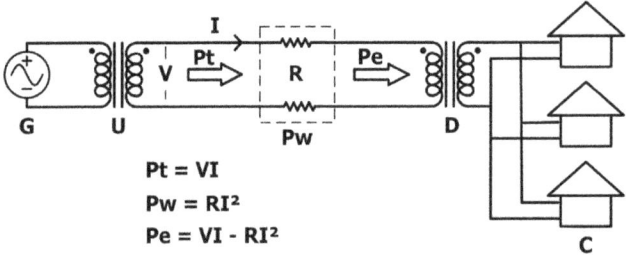

Pt = VI
Pw = RI²
Pe = VI - RI²

A schematic representation of long distance electric power transmission. C=consumers, D=step down transformer, G=generator, I=current in the wires, P_e=power reaching the end of the transmission line, P_t=power entering the transmission line, P_w=power lost in the transmission line, R=total resistance in the wires, V=voltage at the beginning of the transmission line, U=step up transformer.

Electrical energy is distributed as alternating current because AC voltage may be increased or decreased with a transformer. This allows the power to be transmitted through power lines efficiently at high voltage, which reduces the energy lost as heat due to resistance of the wire, and transformed to a lower, safer, voltage for use. Use of a higher voltage leads to significantly more efficient transmission of power. The power losses (P_w) in the wire are a product of the square of the current (I) and the resistance (R) of the wire, described by the formula:

$$P_w = I^2 R.$$

This means that when transmitting a fixed power on a given wire, if the current is halved (i.e. the voltage is doubled), the power loss due to the wire's resistance will be reduced to one quarter.

The power transmitted is equal to the product of the current and the voltage (assuming no phase difference); that is,

$$P_t = IV.$$

Consequently, power transmitted at a higher voltage requires less loss-producing current than for the same power at a lower voltage. Power is often transmitted at hundreds of kilovolts, and transformed to 100 V – 240 V for domestic use.

High voltages have disadvantages, such as the increased insulation required, and generally increased difficulty in their safe handling. In a power plant, energy is generated at a convenient voltage for the design of a generator, and then stepped up to a high voltage for transmission. Near the loads, the transmission voltage is stepped down to the voltages used by equipment. Consumer voltages vary somewhat depending on the country and size of load, but generally motors and lighting are built to use up to a few hundred volts between phases. The voltage delivered to equipment such as lighting and motor loads is standardized, with an allowable range of voltage over which equipment is expected to operate. Standard power utilization voltages and percentage tolerance vary in the different mains power systems found in the world. High-voltage direct-current (HVDC) electric power transmission systems have become more viable as technology has provided efficient means of changing the voltage of DC power. Transmission with high voltage direct current was not feasible in the early days of electric power transmission, as there was then no economically viable way to step down the voltage of DC for end user applications such as lighting incandescent bulbs.

High voltage transmission lines deliver power from electric generation plants over long distances using alternating current. These lines are located in eastern Utah.

Three-phase electrical generation is very common. The simplest way is to use three separate coils in the generator stator, physically offset by an angle of 120° (one-third of a complete 360° phase) to each other. Three current waveforms are produced that are equal in magnitude and 120° out of phase to each other. If coils are added opposite to these (60° spacing), they generate the same phases with reverse polarity and so can be simply wired together. In practice, higher "pole orders" are commonly used. For example, a 12-pole machine would have 36 coils (10° spacing). The advantage is that lower rotational speeds can be used to generate the same frequency. For example, a 2-pole machine running at 3600 rpm and a 12-pole machine running at 600 rpm produce the same frequency; the lower speed is preferable for larger machines. If the load on a three-phase system is balanced equally among the phases, no current flows through the neutral point. Even in the worst-case unbalanced (linear) load, the neutral current will not exceed the highest of the phase currents. Non-linear loads (e.g. the switch-mode power supplies widely used) may require an oversized neutral bus and neutral conductor in the upstream distribution panel to handle harmonics. Harmonics can cause neutral conductor current levels to exceed that of one or all phase conductors.

For three-phase at utilization voltages a four-wire system is often used. When stepping down three-phase, a transformer with a Delta (3-wire) primary and a Star (4-wire, center-earthed) secondary is often used so there is no need for a neutral on the supply side. For smaller customers (just how small varies by country and age of the installation) only a single phase and neutral, or two phases and neutral, are taken to the property. For larger installations all three phases and neutral are taken to the main distribution panel. From the three-phase main panel, both single and three-phase circuits may lead off. Three-wire single-phase systems, with a single center-tapped transformer giving two live conductors, is a common distribution scheme for residential and small commercial buildings in North America. This arrangement is sometimes incorrectly referred to as "two phase". A similar method is used for a different reason on construction sites in the UK. Small power tools and lighting are supposed to be supplied by a local center-tapped transformer with a voltage of 55 V between each power conductor and earth. This significantly reduces the risk of electric shock in the event that one of the live conductors becomes exposed through an equipment fault whilst still allowing a reasonable voltage of 110 V between the two conductors for running the tools.

A third wire, called the bond (or earth) wire, is often connected between non-current-carrying metal enclosures and earth ground. This conductor provides protection from electric shock due to accidental contact of circuit conductors with the metal chassis of portable appliances and tools. Bonding all non-current-carrying metal parts into one complete system ensures there is always a low electrical impedance path to ground sufficient to carry any fault current for as long as it takes for the system to clear the fault. This low impedance path allows the maximum amount of fault current, causing the overcurrent protection device (breakers, fuses) to trip or burn out as quickly as possible, bringing the electrical system to a safe state. All bond wires are bonded to ground at the main service panel, as is the neutral/identified conductor if present.

AC Power Supply Frequencies

The frequency of the electrical system varies by country and sometimes within a country; most electric power is generated at either 50 or 60 Hertz. Some countries have a mixture of 50 Hz and 60 Hz supplies, notably electricity power transmission in Japan. A low frequency eases the design of electric motors, particularly for hoisting, crushing and rolling applications, and commutator-type traction motors for applications such as railways. However, low frequency also causes noticeable flicker in arc lamps and incandescent light bulbs. The use of lower frequencies also provided the advantage of lower impedance losses, which are proportional to frequency. The original Niagara Falls generators were built to produce 25 Hz power, as a compromise between low frequency for traction and heavy induction motors, while still allowing incandescent lighting to operate (although with noticeable flicker). Most of the 25 Hz residential and commercial customers for Niagara Falls power were converted to 60 Hz by the late 1950s, although some 25 Hz industrial customers still existed as of the start of the 21st century. 16.7 Hz power (formerly 16 2/3 Hz) is still used in some European rail systems, such as in Austria, Germany, Norway, Sweden and Switzerland. Off-shore, military, textile industry, marine, aircraft, and spacecraft applications sometimes use 400 Hz, for benefits of reduced weight of apparatus or higher motor speeds. Computer mainframe systems were often powered by 400 Hz or 415 Hz for benefits of ripple reduction while using smaller internal AC to DC conversion units. In any case, the input to the M-G set is the local customary voltage and frequency, variously 200 V (Japan), 208 V, 240 V (North America), 380 V, 400 V or 415 V (Europe), and variously 50 Hz or 60 Hz.

Effects at High Frequencies

A direct current flows uniformly throughout the cross-section of a uniform wire. An alternating current of any frequency is forced away from the wire's center, toward its outer surface. This is because the acceleration of an electric charge in an alternating current produces waves of electromagnetic radiation that cancel the propagation of electricity toward the center of materials with high conductivity.

This phenomenon is called skin effect. At very high frequencies the current no longer flows *in* the wire, but effectively flows *on* the surface of the wire, within a thickness of a few skin depths. The skin depth is the thickness at which the current density is reduced by 63%. Even at relatively low frequencies used for power transmission (50 Hz – 60 Hz), non-uniform distribution of current still occurs in sufficiently thick conductors. For example, the skin depth of a copper conductor is approximately 8.57 mm at 60 Hz, so high current conductors are usually hollow to reduce their mass and cost. Since the current tends to flow in the periphery of conductors, the effective cross-section of the conductor is reduced. This increases the effective AC resistance of the conductor, since resistance is inversely proportional to the cross-sectional area. The AC resistance often is many times higher than the DC resistance, causing a much higher energy loss due to ohmic heating (also called I^2R loss).

Techniques for Reducing AC Resistance

For low to medium frequencies, conductors can be divided into stranded wires, each insulated from one another, with the relative positions of individual strands specially arranged within the conductor bundle. Wire constructed using this technique is called Litz wire. This measure helps to partially mitigate skin effect by forcing more equal current throughout the total cross section of the stranded conductors. Litz wire is used for making high-Q inductors, reducing losses in flexible conductors carrying very high currents at lower frequencies, and in the windings of devices carrying higher radio frequency current (up to hundreds of kilohertz), such as switch-mode power supplies and radio frequency transformers.

Techniques for Reducing Radiation Loss

As written above, an alternating current is made of electric charge under periodic acceleration, which causes radiation of electromagnetic waves. Energy that is radiated is lost. Depending on the frequency, different techniques are used to minimize the loss due to radiation.

Twisted Pairs

At frequencies up to about 1 GHz, pairs of wires are twisted together in a cable, forming a twisted pair. This reduces losses from electromagnetic radiation and inductive coupling. A twisted pair must be used with a balanced signalling system, so that the two wires carry equal but opposite currents. Each wire in a twisted pair radiates a signal, but it is effectively cancelled by radiation from the other wire, resulting in almost no radiation loss.

Coaxial Cables

Coaxial cables are commonly used at audio frequencies and above for convenience. A coaxial cable has a conductive wire inside a conductive tube, separated by a dielectric layer. The current flowing on the surface of the inner conductor is equal and opposite to the current flowing on the inner surface of the outer tube. The electromagnetic field is thus completely contained within the tube, and (ideally) no energy is lost to radiation or coupling outside the tube. Coaxial cables have acceptably small losses for frequencies up to about 5 GHz. For microwave frequencies greater than 5 GHz, the losses (due mainly to the electrical resistance of the central conductor) become too large, making waveguides a more efficient medium for transmitting energy. Coaxial cables with an air rather than solid dielectric are preferred as they transmit power with lower loss.

Waveguides

Waveguides are similar to coaxial cables, as both consist of tubes, with the biggest difference being that the waveguide has no inner conductor. Waveguides can have any arbitrary cross section, but rectangular cross sections are the most common. Because waveguides do not have an inner conductor to carry a return current, waveguides cannot deliver energy by means of an electric current, but rather by means of a *guided* electromagnetic field. Although surface currents do flow on the inner walls of the waveguides, those surface currents do not carry power. Power is carried by the guided electromagnetic fields. The surface currents are set up by the guided electromagnetic fields and have the effect of keeping the fields inside the waveguide and preventing leakage of the fields to the space outside the waveguide. Waveguides have dimensions comparable to the wavelength of the alternating current to be transmitted, so they are only feasible at microwave frequencies. In addition to this mechanical feasibility, electrical resistance of the non-ideal metals forming the walls of the waveguide cause dissipation of power (surface currents flowing on lossy conductors dissipate power). At higher frequencies, the power lost to this dissipation becomes unacceptably large.

Fiber Optics

At frequencies greater than 200 GHz, waveguide dimensions become impractically small, and the ohmic losses in the waveguide walls become large. Instead, fiber optics, which are a form of dielectric waveguides, can be used. For such frequencies, the concepts of voltages and currents are no longer used.

Mathematics of AC Voltages

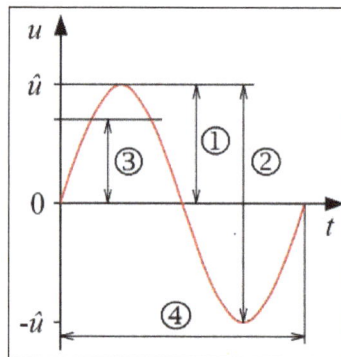

A sinusoidal alternating voltage. 1 = Peak, also amplitude, 2 = Peak-to-peak, 3 = Effective value, 4 = Period.

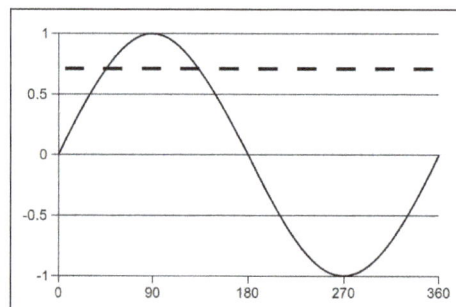

A sine wave, over one cycle (360°). The dashed line represents the root mean square (RMS) value at about 0.707.

Alternating currents are accompanied (or caused) by alternating voltages. An AC voltage v can be described mathematically as a function of time by the following equation:

$$v(t) = V_{peak} \cdot \sin(\omega t),$$

where,

- V_{peak} is the peak voltage (unit: volt).

- ω is the angular frequency (unit: radians per second).

  - The angular frequency is related to the physical frequency, f (unit = hertz), which represents the number of cycles per second, by the equation $\omega = 2\pi f$.

- t is the time (unit: second).

The peak-to-peak value of an AC voltage is defined as the difference between its positive peak and its negative peak. Since the maximum value of $\sin(x)$ is +1 and the minimum value is −1, an AC voltage swings between $+V_{peak}$ and $-V_{peak}$. The peak-to-peak voltage, usually written as V_{pp} or $V_{P\text{-}P}$, is therefore $V_{peak} - (-V_{peak}) = 2V_{peak}$.

Power

The relationship between voltage and the power delivered is:

$$p(t) = \frac{v^2(t)}{R}$$

where, R represents a load resistance.

Rather than using instantaneous power, $p(t)$, it is more practical to use a time averaged power (where the averaging is performed over any integer number of cycles). Therefore, AC voltage is often expressed as a root mean square (RMS) value, written as V_{rms}, because

$$P_{time\ averaged} = \frac{V_{rms}^{\ 2}}{R}.$$

Power Oscillation

$$v(t) = V_{peak} \sin(\omega t)$$

$$i(t) = \frac{v(t)}{R} = \frac{V_{peak}}{R} \sin(\omega t)$$

$$P(t) = v(t)\, i(t) = \frac{(V_{peak})^2}{R} \sin^2(\omega t)$$

Root Mean Square Voltage

Below it is assumed an AC waveform (with no DC component).

The RMS voltage is the square Root of the Mean over one cycle of the Square of the instantaneous voltage.

- For an arbitrary periodic waveform $v(t)$ of period T:

$$V_{rms} = \sqrt{\frac{1}{T}\int_0^T [v(t)]^2 \, dt}.$$

- For a sinusoidal voltage:

$$V_{rms} = \sqrt{\frac{1}{T}\int_0^T [V_{pk}\sin(\omega t + \phi)]^2}$$

$$= V_{pk}\sqrt{\frac{1}{2T}\int_0^T [1 - \cos(2\omega t + 2\phi)]dt}$$

$$= V_{pk}\sqrt{\frac{1}{2T}\int_0^T dt}$$

$$= \frac{V_{pk}}{\sqrt{2}}\, dt$$

where the trigonometric identity $\sin^2 x = \dfrac{1-\cos 2x}{2}$ has been used and the factor $\sqrt{2}$ is called the crest factor, which varies for different waveforms.

- For a triangle waveform centered about zero:

$$V_{rms} = \frac{V_{peak}}{\sqrt{3}}.$$

- For a square waveform centered about zero:

$$V_{rms} = V_{peak}.$$

Example:

To illustrate these concepts, consider a 230V AC mains supply used in many countries around the world. It is so called because its root mean square value is 230V. This means that the time-averaged power delivered is equivalent to the power delivered by a DC voltage of 230V. To determine the peak voltage (amplitude), we can rearrange the above equation to:

$$V_{peak} = \sqrt{2}\, V_{rms}.$$

For 230 V AC, the peak voltage V_{peak} is therefore $230V \times \sqrt{2}$, which is about 325 V. During the course of one cycle the voltage rises from zero to 325 V, falls through zero to -325 V, and returns to zero.

Information Transmission

Alternating current is used to transmit information, as in the cases of telephone and cable television. Information signals are carried over a wide range of AC frequencies. POTS telephone signals have a frequency of about 3 kHz, close to the baseband audio frequency. Cable television and other cable-transmitted information currents may alternate at frequencies of tens to thousands of megahertz. These frequencies are similar to the electromagnetic wave frequencies often used to transmit the same types of information over the air.

DC CIRCUIT

A Direct Current circuit is a circuit that Electric Current flows through in one direction. DC is commonly found in many low-voltage applications, especially where these are powered by Battery. Most electronic circuits require a DC power supply.

A Direct Electric Current flows only when the Electric Circuit is closed, but it stops completely when the circuit is open.

A Switch is a device for making or breaking an Electric Circuit. While the Switch is closed, figure (a), the circuit is closed and the Light Bulb turns On; while the Switch is opened, figure (b), the circuit is open and the Light Bulb turns Off.

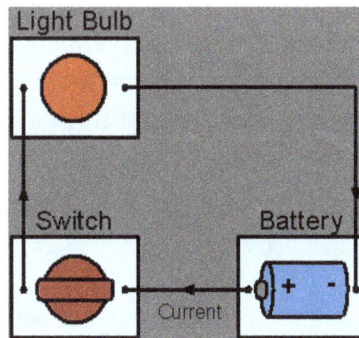

(a): Switch is closed, the circuit is closed and the Light Bulb turns On.

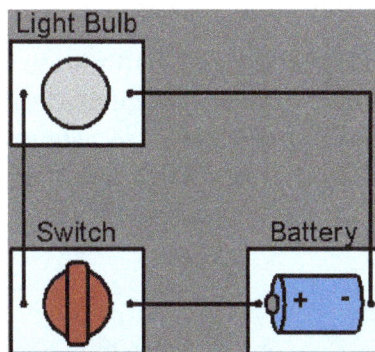

(b): Switch is opened, the circuit is open and the Light Bulb turns Off.

According to Ohm's Law: the Current I in a (ideal) Resistor (or other ohmic device) is proportional to the applied Voltage V and inversely proportional to the Resistance R.

Ohm's Law: I = V / R

In other words, for a fixed Resistance (R), the greater the Voltage (V) across a Resistor, the more the Current (I) flowing through it; for a fixed Voltage across a Resistor, the more the Resistance of the Resistor, the less the Current flowing through it.

In figure, a Resistor is added to the Direct Current Circuit, the total Resistance (R) of the circuit becomes larger but the Power Supply Voltage (V) remains unchanged, therefore, the Current (I) flowing through the circuit is reduced. With less Current flowing through, the Light Bulb becomes dimmer.

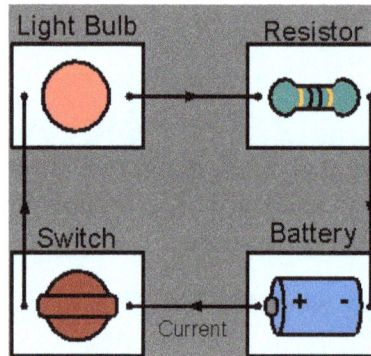

Circuit contains a Battery, a Light Bulb, a Resistor and a Switch.

Direct Current

Direct current (DC) is the unidirectional flow of an electric charge. A battery is a prime example of DC power. Direct current may flow through a conductor such as a wire, but can also flow through semiconductors, insulators, or even through a vacuum as in electron or ion beams. The electric current flows in a constant direction, distinguishing it from alternating current (AC). A term formerly used for this type of current was galvanic current.

The abbreviations *AC* and *DC* are often used to mean simply *alternating* and *direct*, as when they modify *current* or *voltage*.

Direct current may be converted from an alternating current supply by use of a rectifier, which contains electronic elements (usually) or electromechanical elements (historically) that allow current to flow only in one direction. Direct current may be converted into alternating current via an inverter.

Direct current has many uses, from the charging of batteries to large power supplies for electronic systems, motors, and more. Very large quantities of direct-current power are used in production of aluminum and other electrochemical processes. It is also used for some railways, especially in urban areas. High-voltage direct current is used to transmit large amounts of power from remote generation sites or to interconnect alternating current power grids.

Various Definitions

The term *DC* is used to refer to power systems that use only one polarity of voltage or current, and to refer to the constant, zero-frequency, or slowly varying local mean value of a voltage or current. For example, the voltage across a DC voltage source is constant as is the current through a DC current source. The DC solution of an electric circuit is the solution where all voltages and currents are constant. It can be shown that any stationary voltage or current waveform can be decomposed into a sum of a DC component and a zero-mean time-varying component; the DC component is defined to be the expected value, or the average value of the voltage or current over all time.

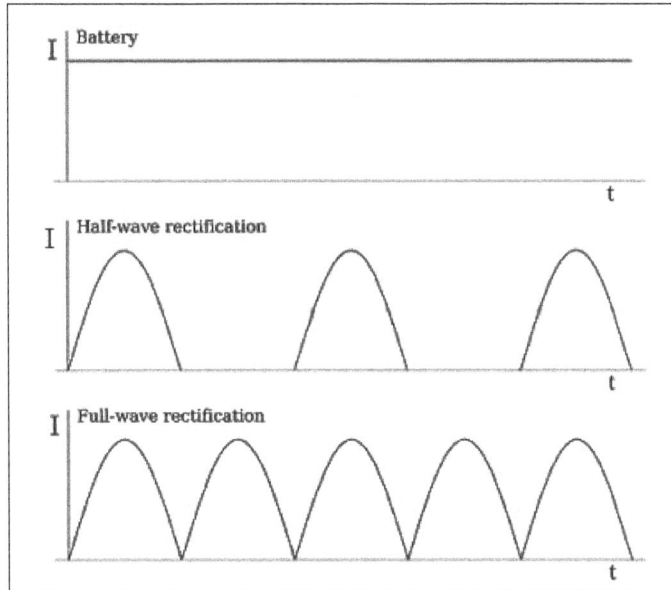

Types of direct current.

Although DC stands for "direct current", DC often refers to "constant polarity". Under this definition, DC voltages can vary in time, as seen in the raw output of a rectifier or the fluctuating voice signal on a telephone line.

Some forms of DC (such as that produced by a voltage regulator) have almost no variations in voltage, but may still have variations in output power and current.

Circuits

A direct current circuit is an electrical circuit that consists of any combination of constant voltage sources, constant current sources, and resistors. In this case, the circuit voltages and currents are independent of time. A particular circuit voltage or current does not depend on the past value of any circuit voltage or current. This implies that the system of equations that represent a DC circuit do not involve integrals or derivatives with respect to time.

If a capacitor or inductor is added to a DC circuit, the resulting circuit is not, strictly speaking, a DC circuit. However, most such circuits have a DC solution. This solution gives the circuit voltages and currents when the circuit is in DC steady state. Such a circuit is represented by a system of differential equations. The solution to these equations usually contain a time varying or transient part as well as constant or steady state part. It is this steady state part that is the DC solution. There are some circuits that do not have a DC solution. Two simple examples are a constant current source connected to a capacitor and a constant voltage source connected to an inductor.

In electronics, it is common to refer to a circuit that is powered by a DC voltage source such as a battery or the output of a DC power supply as a DC circuit even though what is meant is that the circuit is DC powered.

Applications

Domestic and Commercial Buildings

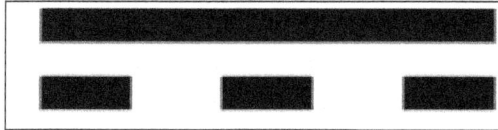

This symbol which can be represented with Unicode character U+2393 (⎓) is found on many electronic devices that either require or produce direct current.

DC is commonly found in many extra-low voltage applications and some low-voltage applications, especially where these are powered by batteries or solar power systems (since both can produce only DC).

Most electronic circuits require a DC power supply.

Domestic DC installations usually have different types of sockets, connectors, switches, and fixtures from those suitable for alternating current. This is mostly due to the lower voltages used, resulting in higher currents to produce the same amount of power.

It is usually important with a DC appliance to observe polarity, unless the device has a diode bridge to correct for this.

EMerge Alliance is the open industry association developing standards of DC power distribution in hybrid houses and commercial buildings.

Automotive

Most automotive applications use DC. An automotive battery provides power for engine starting, lighting, and ignition system. The alternator is an AC device which uses a rectifier to produce DC for battery charging. Most highway passenger vehicles use nominally 12 V systems. Many heavy trucks, farm equipment, or earth moving equipment with Diesel engines use 24 volt systems. In some older vehicles, 6 V was used, such as in the original classic Volkswagen Beetle. At one point a 42 V electrical system was considered for automobiles, but this found little use. To save weight and wire, often the metal frame of the vehicle is connected to one pole of the battery and used as the return conductor in a circuit. Often the negative pole is the chassis "ground" connection, but positive ground may be used in some wheeled or marine vehicles.

Telecommunication

Telephone exchange communication equipment uses standard −48 V DC power supply. The negative polarity is achieved by grounding the positive terminal of power supply system and the battery bank. This is done to prevent electrolysis depositions. Telephone installations have a battery system to ensure power is maintained for subscriber lines during power interruptions.

Other devices may be powered from the telecommunications DC system using a DC-DC converter to provide any convenient voltage.

Many telephones connect to a twisted pair of wires, and use a bias tee to internally separate the AC component of the voltage between the two wires (the audio signal) from the DC component of the voltage between the two wires (used to power the phone).

High-voltage Power Transmission

High-voltage direct current (HVDC) electric power transmission systems use DC for the bulk transmission of electrical power, in contrast with the more common alternating current systems. For long-distance transmission, HVDC systems may be less expensive and suffer lower electrical losses.

Other

Applications using fuel cells (mixing hydrogen and oxygen together with a catalyst to produce electricity and water as byproducts) also produce only DC.

Light aircraft electrical systems are typically 12 V or 24 V DC similar to automobiles.

ELECTRONIC CIRCUIT

An electronic circuit is a complete course of conductors through which current can travel. Circuits provide a path for current to flow. To be a circuit, this path must start and end at the same point. In other words, a circuit must form a loop. An electronic circuit and an electrical circuit has the same definition, but electronic circuits tend to be low voltage circuits.

For example, a simple circuit may include two components: a battery and a lamp. The circuit allows current to flow from the battery to the lamp, through the lamp, then back to the battery. Thus, the circuit forms a complete loop.

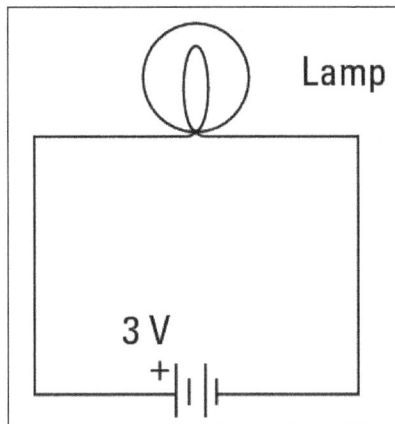

Of course, circuits can be more complex. However, all circuits can be distilled down to three basic elements:

- Voltage source: A voltage source causes current to flow like a battery, for instance.

- Load: The load consumes power; it represents the actual work done by the circuit. Without the load, there's not much point in having a circuit.

The load can be as simple as a single light bulb. In complex circuits, the load is a combination of components, such as resistors, capacitors, transistors, and so on.

- Conductive path: The conductive path provides a route through which current flows. This route begins at the voltage source, travels through the load, and then returns to the voltage source. This path must form a loop from the negative side of the voltage source to the positive side of the voltage source.

The following paragraphs describe a few additional interesting points to keep in mind the nature of basic circuits:

- When a circuit is complete and forms a loop that allows current to flow, the circuit is called a closed circuit. If any part of the circuit is disconnected or disrupted so that a loop is not formed, current cannot flow. In that case, the circuit is called an open circuit.

Open circuit is an oxymoron. After all, the components must form a complete path to be considered a circuit. If the path is open, it isn't a circuit. Therefore, open circuit is most often used to describe a circuit that has become broken, either on purpose (by the use of a switch) or by some error, such as a loose connection or a damaged component.

- Short circuit refers to a circuit that does not have a load. For example, if the lamp is connected to the circuit but a direct connection is present between the battery's negative terminal and its positive terminal, too.

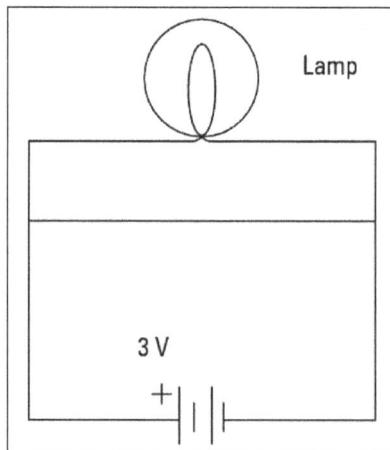

Current in a short circuit can flow at dangerously high levels. Short circuits can damage electronic components, cause a battery to explode, or maybe start a fire.

The short circuit illustrates an important point about electrical circuits: it is possible — common, even — for a circuit to have multiple pathways for current to flow. The current can flow through the lamp as well as through the path that connects the two battery terminals directly.

Current flows everywhere it can. If your circuit has two pathways through which current can flow, the current doesn't choose one over the other; it chooses both. However, not all paths are equal, so current doesn't flow equally through all paths.

For example, current will flow much more easily through the short circuit than it will through the lamp.

Thus, the lamp will not glow because nearly all of the current will bypass the lamp in favor of the easier route through the short circuit. Even so, a small amount of current will flow through the lamp.

How Batteries Work in Electronic Circuits

The easiest way to provide a voltage source for an electronic circuit is to include a battery. There are plenty of other ways to provide voltage, including AC adapters (which you can plug into the wall) and solar cells (which convert sunlight to voltage). However, batteries remain the most practical source of juice for most electronic circuits.

A battery is a device that converts chemical energy into electrical energy in the form of voltage, which in turn can cause current to flow.

A battery works by immersing two plates made of different metals into a special chemical solution called an electrolyte. The metals react with the electrolyte to produce a flow of charges that accumulate on the negative plate, called the anode. The positive plate, called the cathode, is sucked dry of charges. As a result, a voltage is formed between the two plates.

These plates are connected to external terminals to which you can connect a circuit to cause current to flow.

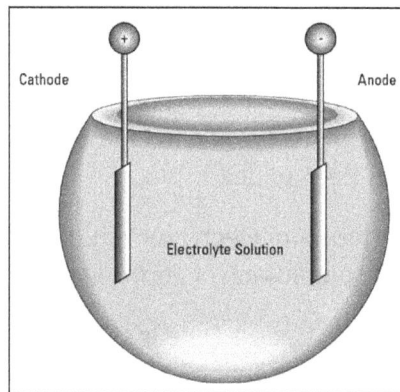

Batteries come in many different shapes and sizes, but for electronics projects, you need concern yourself only with a few standard types of batteries, all of which are available at any grocery, drug, or department store.

Cylindrical batteries come in four standard sizes: AAA, AA, C, and D. Regardless of the size, these batteries provide 1.5 V each; the only difference between the smaller and larger sizes is that the larger batteries can provide more current.

The cathode, or positive terminal, in a cylindrical battery is the end with the metal bump. The flat metal end is the anode, or negative terminal.

The rectangular battery is a 9 V battery. That little rectangular box actually contains six small cells, each about half the size of a AAA cell. The 1.5 volts produced by each of these small cells combine to create a total of 9 volts.

Here are a few other things you should know about batteries:

- Besides AAA, AA, C, D, and 9 V batteries, many other battery sizes are available. Most of those batteries are designed for special applications, such as digital cameras, hearing aids, laptop computers, and so on.

- All batteries contain chemicals that are toxic to you and to the environment. Treat them with care, and dispose of them properly according to your local laws. Don't just throw them in the trash.

- A multimeter can be used to measure the voltage produced by your batteries. Set the multimeter to an appropriate DC voltage range (such as 20 V). Then, touch the red test lead to the positive terminal of the battery and the black test lead to the negative terminal.

The multimeter will tell you the voltage difference between the negative and positive terminals. For cylindrical batteries (AAA, AA, C, or D) it should be about 1.5 V. For 9 V batteries, it should be about 9 V.

- Rechargeable batteries cost more than non-rechargeable batteries but last longer because you can recharge them when they go dead.

- The easiest way to use batteries in an electronic circuit is to use a battery holder, which is a little plastic gadget designed to hold one or more batteries.

- Wonder why they sell AAA, AA, C, and D cells but not A or B? Actually, A cell and B cell batteries exist. However, those sizes never really caught on.

Printed Circuit Board

A printed circuit board (PCB) is an electronic circuit used in devices to provide mechanical support and a pathway to its electronic components. It is made by combining different sheets of non-conductive material, such as fiberglass or plastic, that easily holds copper circuitry.

PCB is also known as printed wiring board (PWB) or etched wiring board (EWB).

A PCB works on the copper films/assembly/circuit that are placed inside of it to provide a pathway for the flow of current. A PCB can hold various electronic components that may be soldered without using visible wires, which facilitates its use.

PCBs are found in nearly every electronic and computing device, including motherboards, network cards and graphics cards to internal circuitry found in hard/CD-ROM drives.

A printed circuit board, or PCB, is found in nearly every type of electronic device. These plastic boards and their embedded components provide the basic technology for everything from computers and mobile phones to smartwatches. The circuit connections on a PCB allow electrical current to be efficiently routed between the miniaturized components on the board, replacing larger devices and bulky wiring.

Functions of a Circuit Board

Depending on the application it's designed for, a PC board may perform a variety of tasks related to computing, communications and data transfer. Aside from the tasks it performs, perhaps the most important function of a circuit board is providing a way to integrate the electronics for a device in a compact space. A PCB allows components to be correctly connected to a power source while being safely insulated. Also, circuit boards are less expensive than other options because they can be designed with digital design tools and manufactured in high volume using factory automation.

Composition of a Circuit Board

A modern circuit board is typically made from layers of different materials. The various layers are fused together through a lamination process. The base material in many boards is fiberglass, which provides a rigid core. A copper foil layer on one or both sides of the board comes next. A chemical process is then used to define copper traces that become conductive paths. These traces take the place of messy wire wrapping found in the point-to-point construction method used for earlier electronics assemblies.

A solder mask layer is added to the circuit board to protect and insulate the copper layer. This plastic layer covers both sides of the board and is frequently green. It's followed by a silkscreen layer with letters, numbers and other identifiers that aid in board assembly. A circuit board's components can be attached to the board in a variety of ways, including soldering. Some attachment methods make use of small holes known as vias that are drilled through the circuit board. Their purpose is to allow electricity to flow from one side of the board to the other.

Basic Circuit Function

A circuit is a loop of conductive material that electricity can travel along. When the loop is closed, electricity can flow uninterrupted from a power source such as a battery through the conductive material and then back to the power source. The design of the circuit is based on the fact that electricity seeks to flow from a higher power voltage, which is a measure of electrical potential, to a lower voltage.

Every circuit is made up of at least four basic elements. The first element is an energy source for either AC or DC power. The second element is a conducting material such as a wire that the energy can move along. This conductive path is known as the track or trace. The third element is the load, which consists of at least one component that drains some of the power to perform a task or operation. The fourth and final element is at least one controller or switch to control the flow of power.

Function of PCB Components

When you insert a load into the closed path of a circuit, the load can use the flow of electrical current to perform an action that requires power. For example, a light emitting diode (LED) component can be made to light up when power flows through the circuit where it's inserted. The load needs to consume energy since a power overload could damage attached components.

The most important components on a circuit board include:

- Battery: Provides power for a circuit, usually through a dual-terminal device that provides a voltage difference between two points in the circuit.

- Capacitor: A battery-like component that can quickly hold or release an electrical charge.

- Diode: Controls electricity on a circuit board by forcing it to flow in one direction.

- Inductor: Stores energy from an electrical current as magnetic energy.

- IC (Integrated Circuit): A chip that may contain many circuits and components in miniaturized form and that typically performs a specific function.

- LED (Light Emitting Diode): A small light used on a circuit board to provide visual feedback.

- Resistor: Regulates the flow of electrical current by providing resistance.

- Switch: Either blocks current or allows it to flow, depending on whether it's closed or open.

- Transistor: A type of switch controlled by electrical signals.

Each of the components on a circuit board performs a specific task or set of tasks that are determined by the overall PCB function. Some of the components such as transistors and capacitors operate directly on electrical currents. They serve as building blocks within more complex components known as integrated circuits.

PCB vs. PCBA

The term PCBA (an acronym for Printed Circuit Board Assembly) is used to describe a circuit board that is completely populated with components attached to the board and connected to the copper traces. It is also referred to as a plug-in assembly. A board that has copper traces but doesn't have components installed is often referred to as a bare board or a printed circuit board.

The design of modern circuit boards allows them to be mass produced at a lower cost than older wire-wrapped boards. After the design phase of a board has been laid out with the aid of specialized computer software, the manufacturing and assembly are – for the most – automated. A PCBA is considered to be finished and ready for use after quality-assurance testing is complete.

Possible Circuit Issues

An open circuit is one that isn't closed due to a broken wire or loose connection. An open circuit won't work because it can't conduct electricity. Although voltage may be available in an open circuit, there is no way for it to flow. In some cases, an open circuit is desired. For example, the switch that's used to turn a light on and off opens and closes the circuit that connects the light to its power source.

Another type of faulty circuit is the short circuit, which can occur when too much power moves through a circuit and damages the conducting material or the power supply. A short circuit can be caused by two points in a circuit connecting when they aren't supposed to, like the two terminals of a power supply being connected with no load component between then to drain some of the current. Shorting out a power supply this way can be dangerous and may even result in fire or an explosion.

Evolution of the Circuit Board

Vacuum tubes and electrical relays performed the basic functions of early computers. The introduction of integrated circuits led to a reduction in both the size and cost of electronic components. Soon circuit boards were developed that contained all the wiring of a device that previously occupied an entire room. These early boards were made from a variety of materials, including Masonite, Bakelite and cardboard, and the connectors consisted of brass wires wrapped around posts.

Beginning in the 1940s, circuit boards became more efficient and cheaper to produce when copper wire replaced brass. Early boards with copper wiring were used on military radios, and by the 1950s, they were being used for consumer devices as well. Soon single-sided boards that contained wiring on only one side evolved to the double-sided and multilayer PCBs that are currently in wide use.

From the 1970s through the 1990s, PCB design became more complex. At the same time, both the physical size and the cost of boards continued to shrink. As boards became denser with attached components, computer-aided design applications (CAD) were developed to aid in their creation. Today there are a variety of tools available for digital PCB design, from free and low-cost options to fully functional, high-priced packages that help with design, manufacturing and testing.

Role of Integrated Circuits

Modern electronics couldn't exist without the integrated circuit, which was introduced in the late 1950s. An IC is a miniaturized collection of circuits and components such as transistors, resistors and diodes assembled on a computer chip to perform a specific function. A single IC chip may contain thousands or even millions of components. The most common types of integrated circuits include logic gates, timers, counters and shift registers.

Besides low-level ICs, there are also more complex microprocessor and microcontroller ICs that have the capability of controlling a computer or another device. Other complex integrated circuits include digital sensors such as accelerometers and gyroscopes that are found in mobile phones and other electronic devices. Like other parts of PCBs, the size of integrated circuits has steadily decreased over the past few decades.

Component Mounting Technologies

Component mounting on early single-sided PCBs used through-hole technology, where a component was attached to one side of the board and fastened through a hole to conductive wire traces on the other side using soldering. At the time it was introduced, through-hole technology was an advancement over point-to-point construction, but holes drilled in the PCB for mounting led to several design issues, especially following the introduction of multilayer boards. Since holes needed to pass through all layers, a large percentage of available real estate on the board was eliminated.

Surface-mount technology (SMT) solved many of the problems caused by through-holes. It became widely used in the 1990s, although it had been introduced several decades earlier. Components were changed to have small pads attached that could be soldered to a circuit board directly instead of through a wire lead. SMT allowed PCB manufacturers to densely package a large number of components on both sides of a PCB. This type of mounting is also easier to manufacture with automation.

SMT mounting did not eliminate the need for holes in circuit boards. Some PCB designs still make use of vias to allow interconnections between components on different layers. However, these holes are not as intrusive as the through-holes used previously for component mounting.

Multilayer Circuit Boards

The most complex electronic devices may include multilayer PCBs. These boards consist of at least three layers of a conductive material such as copper alternating with layers of insulation. Common configurations for multilayer boards include four, six, eight or 10 layers. All the layers must be laminated together to ensure that no air is trapped between the layers. This process is usually done under high temperature and pressure.

The benefits of multilayer PCBs include a higher density of components and circuits in a smaller space. They are used for computers, file servers, GPS technology, health care devices, and satellite and aerospace systems. However, multilayer boards also have some disadvantages. They are more intricate and harder to design and manufacture than single- and double-sided boards, which makes them more expensive. They can also be difficult to repair when something goes wrong within the internal layers of the board.

A Printed Circuit Board, or "PCB," has several major advantages compared to older ways of building electronics. In the past, every component inside of an electronic device was connected with large wires and placed anywhere they could fit. As you can see in this picture of a 1948 television, this method was very "messy," and used a large amount of space. PCBs use a different approach. Components are mounted on a non-conductive board and connected with small pathways, called "traces." Because they are usually designed on a computer, printed circuit boards fit many components in a minimum amount of space.

Parts of a PCB

Vintage 1948 Electronics.

Components Mounted on a PCB.

When looking at a printed circuit board, the traces are easily noticeable. These thin lines are conductive and connect all the components in the circuit. They replace the much larger wires used in the past. A PCB also has many tiny holes. These are drilled exactly where each component needs to be placed. For example, if a microchip is part of the circuit and requires eight connections, the same number of holes will be made on the board. This way, parts of the electronic circuit can be mounted completely flush, without long leads or wires. Again, this saves significant space. The final parts added to a PCB are the components themselves. These are the small electrical devices that must be linked for the unit to work. Common components include microchips, diodes, resisters and switches. The components perform the "work" of the circuit, while the printed circuit board provides the connections.

Creating PCBs

Today, the vast majority of electronic circuits are designed on a computer. This allows electronic engineers to create the perfect arrangement of parts, before making the design permanent. Much like ink on paper, PCBs are literally "printed" when ready. A raw circuit board has two layers, the bottom being non-conductive, and the top being a sheet of metal such as copper. Etch-resistant ink is printed onto the metal layer in the design required. The board is then etched with chemicals. This removes the metal layer, except where the design has been printed. The result: conductive traces remain in the pattern needed for the circuit. Once components are connected to the board with soldering, the PCB is tested and shipped. The entire process is often completely automated, with thousands or millions of the same circuit board being manufactured for use around the world.

Design of a PCB and Final Product.

Modern electronics have come to rely heavily on printed circuit boards. A typical computer may have several PCBs including one complex board called the "motherboard." Individual components can be attached to this board and connected together with a pattern of copper lines covering the board which create pathways for the electricity to travel between components.

Components can be attached to the surface of a circuit board.

Electrical Pathways

The electrical pathways or conductors are made up of two different parts. The first part is the lines themselves and they are called "traces." The second part is called a "land" or "pad." A land is a conductive surface providing a place on which to attach various components, make a connection or provide a test site.

Circuits or pathways conduct electricity and are called "traces."

Single-layer Printed Circuit Boards

Original circuit boards were simply a single-sided circuit board. The board itself is non-conductive but the traces started out as a solid layer of copper and then the spaces in-between were etched away with a chemical bath. Component leads were inserted through holes, soldered from the back and clipped off short.

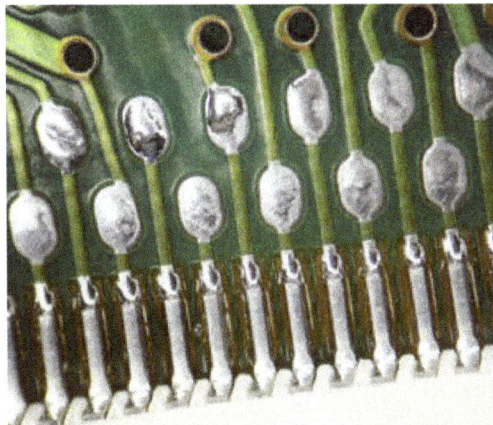

Holes can be used to conduct electricity from one side of the board to another; or to attach wires with solder.

Double-sided Laminate

In an effort to conserve space the double-sided board was invented. By mounting components to the surface of the board rather than using holes, both sides of the circuit board could be used. Holes plated through the board, called a "via," conduct electricity from one side of the board to the other.

Multi-Layer

Further advances came as PCB manufacturers were able to create internal layers that would also conduct electricity allowing even more complex circuit design. This is achieved through a sandwich type of construction. Vias can be designed to only go part way through the board to carry electricity to the internal layers.

Integrated Circuit

A monolithic integrated circuit (also known as IC, microchip, silicon chip, computer chip or chip) is a miniaturized electronic circuit (consisting mainly of semiconductor devices, as well as passive components) that has been manufactured in the surface of a thin substrate of semiconductor material. A hybrid integrated circuit is a miniaturized electronic circuit constructed of individual semiconductor devices, as well as passive components, bonded to a substrate or circuit board. Integrated Circuits can be found in almost every electronic device today. Anything, from a common wristwatch to a personal computer has Integrated Circuits in it. There are circuits that control almost everything, as simple as a temperature control in a common iron or a clock in a microwave oven. Not only does it make electronic items simpler to use, for example, on most microwave ovens now.

In the future, Integrated circuits may even be used for medical purposes. For example, Research has been going on since the late 1980s in which they are trying to develop a computer chip that can be attached to the brain to repair different types of brain damage. With this kind of link, they would be able to repair some kinds of blindness or even memory loss from brain damage.

Only a half-century after their development was initiated, integrated circuits can be found everywhere. Computers, cellular phones, and other digital appliances are now entangled parts of the structure of modern technological societies. In other words, modern computing, communications, manufacturing, and transport systems, including the Internet, all depend on the existence of integrated circuits. Indeed, many scholars believe that the digital revolution that is based on integrated circuits is one of the most significant developments in the history of mankind.

Integrated circuits were made possible by experimental discoveries showing that semiconductor devices could perform the functions of vacuum tubes, and by mid-twentieth-century technology advancements in semiconductor device fabrication. The integration of large numbers of tiny transistors into a small chip was an enormous improvement over the manual assembly of circuits using discrete electronic components. The integrated circuit›s mass production capability, reliability, and building-block approach to circuit design ensured the rapid adoption of standardized ICs in place of designs using discrete transistors.

There are two main advantages of ICs over discrete circuits: cost and performance. Cost is low because the chips, with all their components, are printed as a unit by photolithography and not constructed one transistor at a time. Performance is high, because the components are small, close together, switch quickly, and consume little power. As of 2006, chip areas range from a few square millimeters (mm^2) to around 250 mm^2, with up to 1 million transistors per mm^2.

Advances in Integrated Circuits

Among the most advanced integrated circuits are the microprocessors, that control everything from computers to cellular phones to digital microwave ovens. Digital memory chips are another family of integrated circuit that is crucially important to the modern information society. While the cost of designing and developing a complex integrated circuit is quite high, when spread across typically millions of production units the individual IC cost is minimized. The performance of ICs is high because the small size allows short traces which in turn allows low power logic (such as CMOS) to be used at fast switching speeds.

The integrated circuit from an Intel 8742, an 8-bit microcontroller that includes a CPU running at 12 MHz, 128 bytes of RAM, 2048 bytes of EPROM, and I/O in the same chip.

ICs have consistently migrated to smaller feature sizes over the years, allowing more circuitry to be packed on each chip. This increased capacity per unit area can be used to decrease cost and/or increase functionality. Moore's law, in its modern interpretation, states that the number of transistors in an integrated circuit doubles every two years. In general, as the feature size shrinks, almost everything improves—the cost-per-unit and the switching power consumption go down, and the speed goes up. However, ICs with nanometer-scale devices are not without their problems, principal among which is leakage current, although these problems are not insurmountable and will likely be improved by the introduction of high-k dielectrics. Since these speed and power consumption gains are apparent to the end user, there is fierce competition among manufacturers to use finer geometries. This process, and the expected progress over the next few years, is well described by the International Technology Roadmap for Semiconductors (ITRS).

Classification

Integrated circuits can be classified into analog, digital and mixed signal (both analog and digital on the same chip).

A CMOS 4000 IC.

Digital integrated circuits can contain anything from one to millions of logic gates, flip-flops, multiplexers, and other circuits in a few square millimeters. The small size of these circuits allows high speed, low power dissipation, and reduced manufacturing cost compared with board-level integration. These digital ICs, typically microprocessors, digital signal processors (DSPs), and microcontrollers work using binary mathematics to process "one" and "zero" signals.

Analog ICs, such as sensors, power-management circuits, and operational amplifiers work by processing continuous signals. They perform functions like amplification, active filtering, demodulation, mixing, etc. Analog ICs ease the burden on circuit designers by having expertly designed analog circuits available instead of designing a difficult analog circuit from scratch.

ICs can also combine analog and digital circuits on a single chip to create functions such as analog-to-digital converters and digital-to-analog converters. Such circuits offer smaller size and lower cost, but must carefully account for signal interference.

Manufacture

Fabrication

The semiconductors of the periodic table of the chemical elements were identified as the most likely materials for a solid state vacuum tube by researchers like William Shockley at Bell Laboratories starting in the 1930s. Starting with copper oxide, proceeding to germanium, then silicon, the materials were systematically studied in the 1940s and 1950s. Today, silicon monocrystals are the main substrate used for integrated circuits (ICs) although some III-V compounds of the periodic table such as gallium arsenide are used for specialized applications like LEDs, lasers, and the highest-speed integrated circuits. It took decades to perfect methods of creating crystals without defects in the crystalline structure of the semiconducting material.

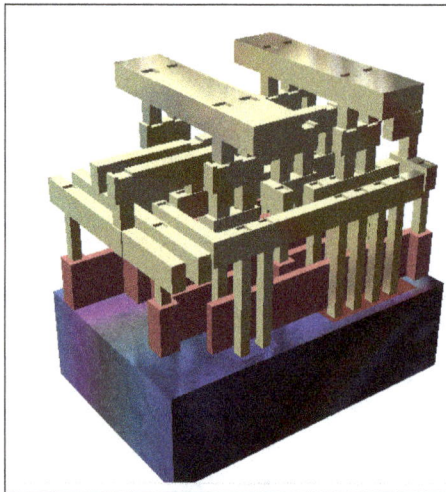

Rendering of a Small Standard Cell with Three Metal Layers (Dielectric has been removed).

Semiconductor ICs are fabricated in a layer process which includes these key process steps:

- Imaging
- Deposition
- Etching

The main process steps are supplemented by doping, cleaning and planarisation steps.

Mono-crystal silicon wafers (or for special applications, silicon on sapphire or gallium arsenide wafers) are used as the substrate. Photolithography is used to mark different areas of the substrate to be doped or to have polysilicon, insulators or metal (typically aluminum) tracks deposited on them.

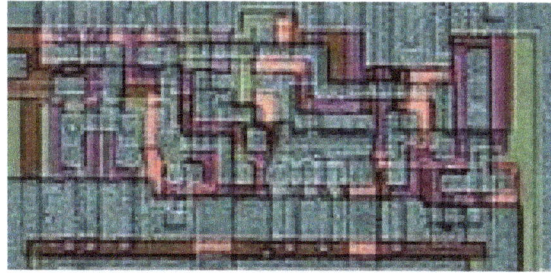

Criss-crossing geometry of the layers of an IC.

- For a CMOS process, for example, a transistor is formed by the criss-crossing intersection of striped layers. The stripes can be monocrystalline substrate, doped layers, perhaps insulator layers or polysilicon layers. Some etched vias to the doped layers might interconnect layers with metal conducting tracks.

- The criss-crossed checkerboard-like transistors are the most common part of the circuit, each checker forming a transistor.

- Resistive structures, meandering stripes of varying lengths, form the loads on the circuit. The ratio of the length of the resistive structure to its width, combined with its sheet resistivity determines the resistance.

- Capacitive structures, in form very much like the parallel conducting plates of a traditional electrical capacitor, are formed according to the area of the "plates," with insulating material between the plates. Owing to limitations in size, only very small capacitances can be created on an IC.

- More rarely, inductive structures can be simulated by gyrators.

- Since a CMOS device only draws current on the transition between logic states, CMOS devices consume much less current than bipolar devices.

A (random access memory) is the most regular type of integrated circuit; the highest density devices are thus memories; but even a microprocessor will have memory on the chip. Although the structures are intricate—with widths which have been shrinking for decades—the layers remain much thinner than the device widths. The layers of material are fabricated much like a photographic process, although light waves in the visible spectrum cannot be used to "expose" a layer of material, as they would be too large for the features. Thus photons of higher frequencies (typically ultraviolet) are used to create the patterns for each layer. Because each feature is so small, electron microscopes are essential tools for a process engineer who might be debugging a fabrication process.

Each device is tested before packaging using very expensive automated test equipment (ATE), a process known as wafer testing, or wafer probing. The wafer is then cut into small rectangles called dice. Each good die (N.B. die is the singular form of dice, although dies is also used as the plural) is then connected into a package using aluminum (or gold) wires which are welded to pads, usually found around the edge of the die. After packaging, the devices go through final test on the same or similar ATE used during wafer probing. Test cost can account for over 25 percent of the cost of fabrication on lower cost products, but can be negligible on low yielding, larger, and/or higher cost devices.

As of 2005, a fabrication facility (commonly known as a semiconductor fab) costs over a billion US Dollars to construct, because much of the operation is automated. The most advanced processes employ the following specifications:

- The wafers are up to 300 mm in diameter (wider than a common dinner plate).

- Use of 90 nanometer or smaller chip manufacturing process. Intel, IBM, and AMD are using 90 nanometers for their CPU chips, and Intel has started using a 65 nanometer process.

- Copper interconnects where copper wiring replaces aluminum for interconnects.

- Low-K dielectric insulators.

- Silicon on insulator (SOI).

- Strained silicon in a process used by IBM known as Strained silicon directly on insulator (SSDOI).

Packaging

The earliest integrated circuits were packaged in ceramic flat packs, which continued to be used by the military for their reliability and small size for many years. Commercial circuit packaging quickly moved to the dual in-line package (DIP), first in ceramic and later in plastic. In the 1980s, pin counts of VLSI circuits exceeded the practical limit for DIP packaging, leading to pin grid array (PGA) and leadless chip carrier (LCC) packages. Surface mount packaging appeared in the early 1980s and became popular in the late 1980s, using finer lead pitch with leads formed as either gull-wing or J-lead, as exemplified by Small-Outline Integrated Circuit. A carrier which occupies an area about 30 percent − 50 percent less than an equivalent DIP, with a typical thickness that is 70 percent less. This package has "gull wing" leads protruding from the two long sides and a lead spacing of 0.050 inches.

Small-Outline Integrated Circuit (SOIC) and PLCC packages. In the late 1990s, PQFP and TSOP packages became the most common for high pin count devices, though PGA packages are still often used for high-end microprocessors. Intel and AMD are currently transitioning from PGA packages on high-end microprocessors to land grid array (LGA) packages.

Ball grid array (BGA) packages have existed since the 1970s.

Traces out of the die, through the package, and into the printed circuit board have very different electrical properties, compared to on-chip signals. They require special design techniques and need much more electric power than signals confined to the chip itself.

When multiple die are put in one package, it is called SiP, for System In Package. When multiple die are combined on a small substrate, often ceramic, it's called a MCM, or Multi-Chip Module. The boundary between a big MCM and a small printed circuit board is sometimes fuzzy.

Mixed-signal Integrated Circuit

A mixed-signal integrated circuit is any integrated circuit that has both analog circuits and digital circuits on a single semiconductor die. In real-life applications mixed-signal designs are everywhere, for example, smart mobile phones. Mixed-signal ICs also process both analog and digital signals together. For example, an analog-to-digital converter is a mixed-signal circuit. Mixed-signal circuits or systems are typically cost-effective solutions for building any modern consumer electronics applications.

An analog-mixed-signal system-on-a-chip (AMS-SoC) can be a combination of analog circuits, digital circuits, intrinsic mixed-signal circuits (like ADC), and embedded software.

Integrated circuits (ICs) are generally classified as digital (e.g. a microprocessor) or analog (e.g. an operational amplifier). Mixed-signal ICs are chips that contain both digital and analog circuits on the same chip. This category of chip has grown dramatically with the increased use of 3G cell phones and other portable technologies.

Mixed-signal ICs are often used to convert analog signals to digital signals so that digital devices can process them. For example, mixed-signal ICs are essential components for FM tuners in digital products such as media players, which have digital amplifiers. Any analog signal (such as an FM radio transmission, a light wave or a sound) can be digitized using a very basic analog-to-digital converter, and the smallest and most energy efficient of these would be in the form of mixed-signal ICs.

Mixed-signal ICs are more difficult to design and manufacture than analog-only or digital-only integrated circuits. For example, an efficient mixed-signal IC would have its digital and analog components share a common power supply. However, analog and digital components have very different power needs and consumption characteristics that make this a non-trivial goal in chip design.

Examples:

Typically, mixed-signal chips perform some whole function or sub-function in a larger assembly such as the radio subsystem of a cell phone, or the read data path and laser sled control logic of a DVD player. They often contain an entire system-on-a-chip.

Examples of mixed-signal integrated circuits include data converters using delta-sigma modulation, analog-to-digital converter/digital-to-analog converter using error detection and correction, and digital radio chips. Digitally controlled sound chips are also mixed-signal circuits. With the advent of cellular technology and network technology this category now includes cellular telephone, software radio, LAN and WAN router integrated circuits.

Because of the use of both digital signal processing and analog circuitry, mixed-signal ICs are usually designed for a very specific purpose and their design requires a high level of expertise and careful use of computer aided design (CAD) tools. Automated testing of the finished chips can also be challenging. Teradyne, Keysight, and Texas Instruments are the major suppliers of the test equipment for mixed-signal chips.

The particular challenges of mixed signal include:

- CMOS technology is usually optimal for digital performance and scaling while bipolar transistors are usually optimal for analog performance, yet until the last decade it has been difficult to either combine these cost-effectively or to design both analog and digital in a single technology without serious performance compromises. The advent of technologies like high performance CMOS, BiCMOS, CMOS SOI and SiGe have removed many of the compromises that previously had to be made.

- Testing functional operation of mixed-signal ICs remains complex, expensive and often a "one-off" implementation task.

- Systematic design methodologies comparable to digital design methods are far more primitive in the analog and mixed-signal arena. Analog circuit design can not generally be automated to nearly the extent that digital circuit design can. Combining the two technologies multiplies this complication.

- Fast-changing digital signals send noise to sensitive analog inputs. One path for this noise is substrate coupling. A variety of techniques are used to attempt to block or cancel this noise coupling, such as fully differential amplifiers, P+ guard-rings, differential topology, on-chip decoupling, and triple-well isolation.

Commercial examples:

- ICsense,

- AnSem,

- Atari POKEY,

- MOS Technology SID,

- PSoC – Cypress PSoC Programmable System on Chip,

- Silego Technology Inc.,

- System to ASIC,

- Texas Instruments' MSP430,

- Triad Semiconductor,

- Wolfson Microelectronics.

Most modern radio and communications use mixed signal circuits.

Current Source

A current source is an electronic circuit that delivers or absorbs an electric current which is independent of the voltage across it.

A current source is the dual of a voltage source. The term *current sink* is sometimes used for sources fed from a negative voltage supply. Figure shows the schematic symbol for an ideal current source driving a resistive load. There are two types. An *independent current source* (or sink) delivers a constant current. A *dependent current source* delivers a current which is proportional to some other voltage or current in the circuit.

An ideal current source generates a current that is independent of the voltage changes across it. An ideal current source is a mathematical model, which real devices can approach very closely. If the current through an ideal current source can be specified independently of any other variable in a circuit, it is called an *independent* current source. Conversely, if the current through an ideal current source is determined by some other voltage or current in a circuit, it is called a dependent or controlled current source.

The internal resistance of an ideal current source is infinite. An independent current source with zero current is identical to an ideal open circuit. The voltage across an ideal current source is completely determined by the circuit it is connected to. When connected to a short circuit, there is zero voltage and thus zero power delivered. When connected to a load resistance, the voltage across the source approaches infinity as the load resistance approaches infinity (an open circuit).

No physical current source is ideal. For example, no physical current source can operate when applied to an open circuit. There are two characteristics that define a current source in real life. One is its internal resistance and the other is its compliance voltage. The compliance voltage is the maximum voltage that the current source can supply to a load. Over a given load range, it is possible for some types of real current sources to exhibit nearly infinite internal resistance. However, when the current source reaches its compliance voltage, it abruptly stops being a current source.

In circuit analysis, a current source having finite internal resistance is modeled by placing the value of that resistance across an ideal current source (the Norton equivalent circuit). However, this model is only useful when a current source is operating within its compliance voltage.

Implementations

Passive Current Source

The simplest non-ideal current source consists of a voltage source in series with a resistor. The amount of current available from such a source is given by the ratio of the voltage across the voltage source to the resistance of the resistor (Ohm's law; $I = V/R$). This value of current will only be delivered to a load with zero voltage drop across its terminals (a short circuit, an uncharged capacitor, a charged inductor, a virtual ground circuit, etc.) The current delivered to a load with nonzero voltage (drop) across its terminals (a linear or nonlinear resistor with a finite resistance, a charged capacitor, an uncharged inductor, a voltage source, etc.) will always be different. It is given by the ratio of the voltage drop across the resistor (the difference between the exciting voltage and the voltage across the load) to its resistance. For a nearly ideal current source, the value of the resistor should be very large but this implies that, for a specified current, the voltage source must be very large (in the limit as the resistance and the voltage go to infinity, the current source will become ideal and the current will not depend at all on the voltage across the load). Thus, efficiency is low (due to power loss in the resistor) and it is usually impractical to construct a 'good' current source this way. Nonetheless, it is often the case that such a circuit will provide adequate performance when the specified current and load resistance are small. For example, a 5 V voltage source in series with a 4.7 kilohm resistor will provide an *approximately* constant current of 1 mA ± 5% to a load resistance in the range of 50 to 450 ohm.

A Van de Graaff generator is an example of such a high voltage current source. It behaves as an almost constant current source because of its very high output voltage coupled with its very high output resistance and so it supplies the same few microamperes at any output voltage up to hundreds of thousands of volts (or even tens of megavolts) for large laboratory versions.

Active Current Sources without Negative Feedback

In these circuits the output current is not monitored and controlled by means of negative feedback.

Current-stable Nonlinear Implementation

They are implemented by active electronic components (transistors) having current-stable nonlinear output characteristic when driven by steady input quantity (current or voltage). These circuits behave as dynamic resistors changing their present resistance to compensate current variations. For example, if the load increases its resistance, the transistor decreases its present output resistance (and *vice versa*) to keep up a constant total resistance in the circuit.

Active current sources have many important applications in electronic circuits. They are often used in place of ohmic resistors in analog integrated circuits (e.g., a differential amplifier) to generate a current that depends slightly on the voltage across the load.

The common emitter configuration driven by a constant input current or voltage and common source (common cathode) driven by a constant voltage naturally behave as current sources (or sinks) because the output impedance of these devices is naturally high. The output part of the simple current mirror is an example of such a current source widely used in integrated circuits. The common base, common gate and common grid configurations can serve as constant current sources as well.

A JFET can be made to act as a current source by tying its gate to its source. The current then flowing is the I_{DSS} of the FET. These can be purchased with this connection already made and in this case the devices are called current regulator diodes or constant current diodes or current limiting diodes (CLD). An enhancement mode N channel MOSFET can be used in the circuits listed below.

Following Voltage Implementation

Voltage Compensation Implementation

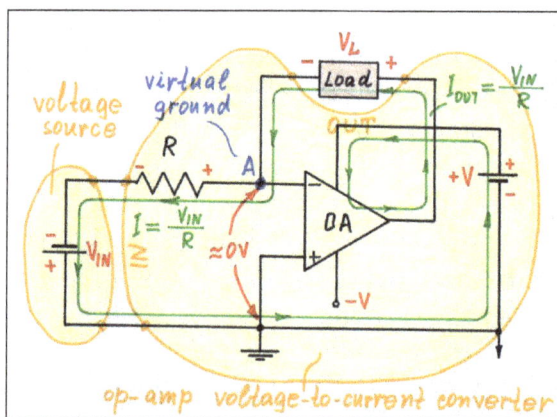

In an op-amp *voltage-controlled current source* the op-amp compensates the voltage drop across the load by adding the same voltage to the exciting input voltage.

The simple resistor passive current source is ideal only when the voltage across it is 0; so voltage compensation by applying parallel negative feedback might be considered to improve the source. Operational amplifiers with feedback effectively work to minimise the voltage across their inputs.

This results in making the inverting input a virtual ground, with the current running through the feedback, or load, and the passive current source. The input voltage source, the resistor, and the op-amp constitutes an "ideal" current source with value, $I_{OUT} = V_{IN}/R$. The op-amp voltage-to-current converter in figure, a transimpedance amplifier and an op-amp inverting amplifier are typical implementations of this idea. The floating load is a serious disadvantage of this circuit solution.

Current Compensation Implementation

A typical example are Howland current source and its derivative Deboo integrator. In the last example, the Howland current source consists of an input voltage source, V_{IN}, a positive resistor, R, a load (the capacitor, C, acting as impedance Z) and a negative impedance converter INIC ($R_1 = R_2 = R_3 = R$ and the op-amp). The input voltage source and the resistor R constitute an imperfect current source passing current, I_R through the load. The INIC acts as a second current source passing "helping" current, I_{-R}, through the load. As a result, the total current flowing through the load is constant and the circuit impedance seen by the input source is increased. However the Howland current source isn't widely used because it requires the four resistors to be perfectly matched, and its impedance drops at high frequencies.

The grounded load is an advantage of this circuit solution.

Current Sources with Negative Feedback

They are implemented as a voltage follower with series negative feedback driven by a constant input voltage source (i.e., a *negative feedback voltage stabilizer*). The voltage follower is loaded by a constant (current sensing) resistor acting as a simple current-to-voltage converter connected in the feedback loop. The external load of this current source is connected somewhere in the path of the current supplying the current sensing resistor but out of the feedback loop.

The voltage follower adjusts its output current I_{OUT} flowing through the load so that to make the voltage drop $V_R = I_{OUT}R$ across the current sensing resistor R equal to the constant input voltage V_{IN}. Thus the voltage stabilizer keeps up a constant voltage drop across a constant resistor; so, a constant current $I_{OUT} = V_R/R = V_{IN}/R$ flows through the resistor and respectively through the load.

If the input voltage varies, this arrangement will act as a voltage-to-current converter (voltage-controlled current source, VCCS); it can be thought as a reversed (by means of negative feedback) current-to-voltage converter. The resistance R determines the transfer ratio (transconductance).

Current sources implemented as circuits with series negative feedback have the disadvantage that the voltage drop across the current sensing resistor decreases the maximal voltage across the load (the *compliance voltage*).

Simple Transistor Current Sources

Constant Current Diode

The simplest constant-current source or sink is formed from one component: a JFET with its gate attached to its source. Once the drain-source voltage reaches a certain minimum value, the JFET enters saturation where current is approximately constant. This configuration is known as a constant-current diode, as it behaves much like a dual to the constant voltage diode (Zener diode) used in simple voltage sources.

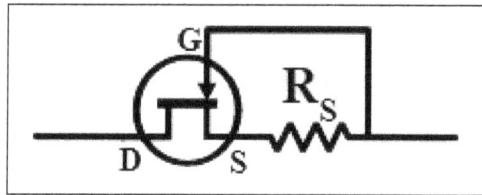

The internal structure of a current limiting diode.

Due to the large variability in saturation current of JFETs, it is common to also include a source resistor which allows the current to be tuned down to a desired value.

Zener Diode Current Source

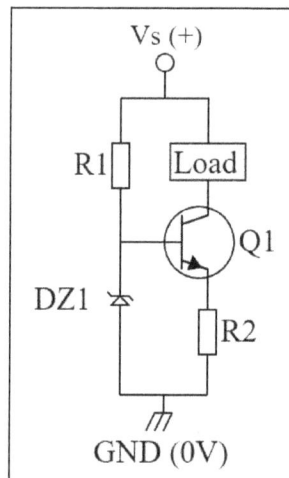

Typical BJT constant current source with negative feedback.

In this bipolar junction transistor (BJT) implementation of the general idea above, a *Zener voltage stabilizer* (R1 and DZ1) drives an *emitter follower* (Q1) loaded by a *constant emitter resistor* (R2) sensing the load current. The external (floating) load of this current source is connected to the collector so that almost the same current flows through it and the emitter resistor (they can be thought of as connected in series). The transistor, Q1, adjusts the output (collector) current so as to keep the voltage drop across the constant emitter resistor, R2, almost equal to the relatively constant voltage drop across the Zener diode, DZ1. As a result, the output current is almost constant even if the load resistance and/or voltage vary. The operation of the circuit is considered in details below.

A Zener diode, when reverse biased (as shown in the circuit) has a constant voltage drop across it irrespective of the current flowing through it. Thus, as long as the Zener current (I_z) is above a certain level (called holding current), the voltage across the Zener diode (V_z) will be constant. Resistor, R1, supplies the Zener current and the base current (I_B) of NPN transistor (Q1). The constant Zener voltage is applied across the base of Q1 and emitter resistor, R2.

Voltage across R2 (V_{R2}) is given by $V_Z - V_{BE}$, where V_{BE} is the base-emitter drop of Q1. The emitter current of Q1 which is also the current through R2 is given by,

$$I_{R2}(= I_E = I_C) = \frac{V_{R2}}{R_{R2}} = \frac{V_Z - V_{BE}}{R_{R2}}.$$

Since V_Z is constant and V_{BE} is also (approximately) constant for a given temperature, it follows that V_{R2} is constant and hence I_E is also constant. Due to transistor action, emitter current, I_E, is very nearly equal to the collector current, I_C, of the transistor (which in turn, is the current through the load). Thus, the load current is constant (neglecting the output resistance of the transistor due to the Early effect) and the circuit operates as a constant current source. As long as the temperature remains constant (or doesn't vary much), the load current will be independent of the supply voltage, R1 and the transistor's gain. R2 allows the load current to be set at any desirable value and is calculated by,

$$R_{R2} = \frac{V_Z - V_{BE}}{I_{R2}}$$

where V_{BE} is typically 0.65 V for a silicon device.

(I_{R2} is also the emitter current and is assumed to be the same as the collector or required load current, provided h_{FE} is sufficiently large). Resistance, R_{R1}, at resistor, R1, is calculated as,

$$R_{R1} = \frac{V_S - V_Z}{I_Z + K \cdot I_B}$$

where K = 1.2 to 2 (so that R_{R1} is low enough to ensure adequate I_B),

$$I_B = \frac{I_C}{h_{FE,min}}$$

and $h_{FE,min}$ is the lowest acceptable current gain for the particular transistor type being used.

LED Current Source

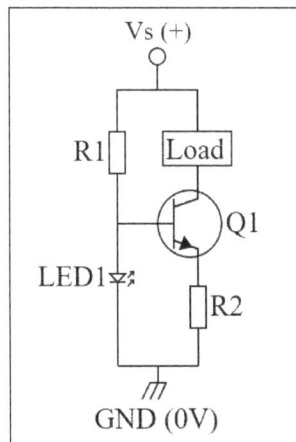

Typical constant current source (CCS) using LED instead of Zener diode.

The Zener diode can be replaced by any other diode; e.g., a light-emitting diode LED1 as shown in figure. The LED voltage drop (V_D) is now used to derive the constant voltage and also has the additional advantage of tracking (compensating) V_{BE} changes due to temperature. R_{R2} is calculated as,

$$R_{R2} = \frac{V_D - V_{BE}}{I_{R2}}$$

and R_1 as,

$$R_{R1} = \frac{V_S - V_D}{I_D + K \cdot I_B} \text{ , where } I_D \text{ is the LED current.}$$

Transistor Current Source with Diode Compensation

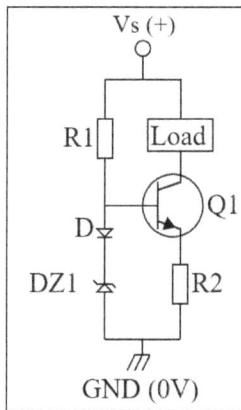

Typical constant current source (CCS) with diode compensation.

Temperature changes will change the output current delivered by the circuit of figure because V_{BE} is sensitive to temperature. Temperature dependence can be compensated using the circuit of figure that includes a standard diode, D, (of the same semiconductor material as the transistor) in series with the Zener diode as shown in the image on the left. The diode drop (V_D) tracks the V_{BE} changes due to temperature and thus significantly counteracts temperature dependence of the CCS.

Resistance R_2 is now calculated as,

$$R_2 = \frac{V_Z + V_D - V_{BE}}{I_{R2}}$$

Since $V_D = V_{BE} = 0.65$ V,

$$R_2 = \frac{V_Z}{I_{R2}}$$

(In practice, V_D is never exactly equal to V_{BE} and hence it only suppresses the change in V_{BE} rather than nulling it out.)

R_1 is calculated as,

$$R_1 = \frac{V_S - V_Z - V_D}{I_Z + K \cdot I_B}$$

(the compensating diode's forward voltage drop, V_D, appears in the equation and is typically 0.65 V for silicon devices.)

Current Mirror with Emitter Degeneration

Series negative feedback is also used in the two-transistor current mirror with emitter degeneration. Negative feedback is a basic feature in some current mirrors using multiple transistors, such as the Widlar current source and the Wilson current source.

Constant Current Source with Thermal Compensation

One limitation with the circuits, thermal compensation is imperfect. In bipolar transistors, as the junction temperature increases the V_{be} drop (voltage drop from base to emitter) decreases. In the two previous circuits, a decrease in V_{be} will cause an increase in voltage across the emitter resistor, which in turn will cause an increase in collector current drawn through the load. The end result is that the amount of 'constant' current supplied is at least somewhat dependent on temperature. This effect is mitigated to a large extent, but not completely, by corresponding voltage drops for the diode, D1 and the LED, LED1. If the power dissipation in the active device of the CCS is not small and/or insufficient emitter degeneration is used, this can become a non-trivial issue.

The LED has 1 V across it driving the base of the transistor. At room temperature there is about 0.6 V drop across the V_{be} junction and hence 0.4 V across the emitter resistor, giving an approximate collector (load) current of $0.4/R_e$ amps. Now imagine that the power dissipation in the transistor causes it to heat up. This causes the V_{be} drop (which was 0.6 V at room temperature) to drop to, say, 0.2 V. Now the voltage across the emitter resistor is 0.8 V, twice what it was before the warmup. This means that the collector (load) current is now twice the design value! This is an extreme example of course, but serves to illustrate the issue.

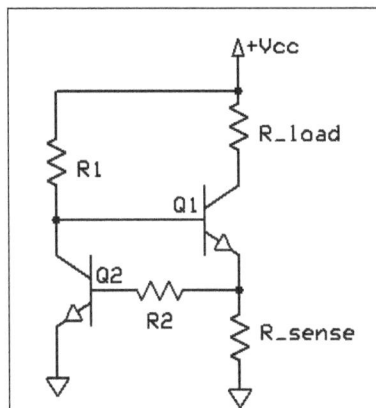

Current Limiter with NPN Transistors.

The circuit to the left overcomes the thermal problem. To see how the circuit works, assume the voltage has just been applied at V+. Current runs through R1 to the base of Q1, turning it on and causing current to begin to flow through the load into the collector of Q1. This same load current then flows out of Q1's emitter and consequently through R_{sense} to ground. When this current through R_{sense} to ground is sufficient to cause a voltage drop that is equal to the V_{be} drop of Q2, Q2 begins to turn on. As Q2 turns on it pulls more current through its collector resistor, R1, which diverts some of the injected current in the base of Q1, causing Q1 to conduct less current through the load. This creates a negative feedback loop within the circuit, which keeps the voltage at Q1's emitter almost exactly equal to the V_{be} drop of Q2. Since Q2 is dissipating very little power compared to Q1 (since all the load current goes through Q1, not Q2), Q2 will not heat up any significant amount and the reference (current setting) voltage across R_{sense} will remain steady at ~0.6 V, or one diode drop above ground, regardless of the thermal changes in the V_{be} drop of Q1. The circuit is still sensitive to changes in the ambient temperature in which the device operates as the BE voltage drop in Q2 varies slightly with temperature.

Op-amp Current Sources

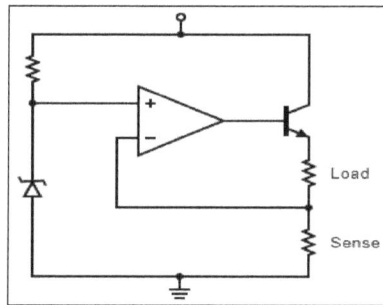

Typical op-amp current source.

The simple transistor current source can be improved by inserting the base-emitter junction of the transistor in the feedback loop of an op-amp. Now the op-amp increases its output voltage to compensate for the V_{BE} drop. The circuit is actually a buffered non-inverting amplifier driven by a constant input voltage. It keeps up this constant voltage across the constant sense resistor. As a result, the current flowing through the load is constant as well; it is exactly the Zener voltage divided by the sense resistor. The load can be connected either in the emitter or in the collector but in both the cases it is floating as in all the circuits above. The transistor is not needed if the required current doesn't exceed the sourcing ability of the op-amp.

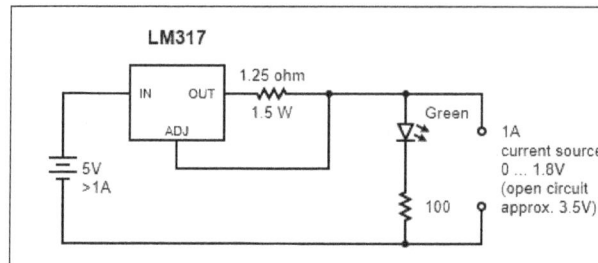

Constant current source using the LM317 voltage regulator.

Voltage Regulator Current Sources

The general negative feedback arrangement can be implemented by an IC voltage regulator. As with the bare emitter follower and the precise op-amp follower above, it keeps up a constant voltage drop (1.25 V) across a constant resistor (1.25 Ω); so, a constant current (1 A) flows through the resistor and the load. The LED is on when the voltage across the load exceeds 1.8 V (the indicator circuit introduces some error). The grounded load is an important advantage of this solution.

Curpistor Tubes

Nitrogen-filled glass tubes with two electrodes and a calibrated Becquerel (fissions per second) amount of 226_{Ra} offer a constant number of charge carriers per second for conduction, which determines the maximum current the tube can pass over a voltage range from 25 to 500 V.

Current and Voltage Source Comparison

Most sources of electrical energy (mains electricity, a battery, etc.) are best modeled as voltage sources. Such sources provide constant voltage, which means that as long as the current drawn from the source is within the source's capabilities, its output voltage stays constant. An ideal voltage source provides no energy when it is loaded by an open circuit (i.e., an infinite impedance), but approaches infinite power and current when the load resistance approaches zero (a short circuit). Such a theoretical device would have a zero ohm output impedance in series with the source. A real-world voltage source has a very low, but non-zero output impedance: often much less than 1 ohm.

Conversely, a current source provides a constant current, as long as the load connected to the source terminals has sufficiently low impedance. An ideal current source would provide no energy to a short circuit and approach infinite energy and voltage as the load resistance approaches infinity (an open circuit). An *ideal* current source has an infinite output impedance in parallel with the source. A *real-world* current source has a very high, but finite output impedance. In the case of transistor current sources, impedances of a few megohms (at DC) are typical.

An *ideal* current source cannot be connected to an *ideal* open circuit because this would create the paradox of running a constant, non-zero current (from the current source) through an element with a defined zero current (the open circuit). Also, a current source should not be connected to another current source if their currents differ but this arrangement is frequently used (e.g., in amplifying stages with dynamic load, CMOS circuits, etc.)

Similarly, an *ideal* voltage source cannot be connected to an *ideal* short circuit (R = 0), since this would result a similar paradox of finite non-zero voltage across an element with defined zero voltage (the short circuit). Also, a voltage source should not be connected to another voltage source if their voltages differ but again this arrangement is frequently used (e.g., in common base and differential amplifying stages).

Contrary, current and voltage sources can be connected to each other without any problems, and this technique is widely used in circuitry (e.g., in cascode circuits, differential amplifier stages with common emitter current source, etc.)

Because no ideal sources of either variety exist (all real-world examples have finite and non-zero source impedance), any current source can be considered as a voltage source with the *same* source impedance and vice versa. These concepts are dealt with by Norton's and Thévenin's theorems.

Charging of capacitor by constant current source and by voltage source is different. Linearity is maintained for constant current source charging of capacitor with time, whereas voltage source charging of capacitor is exponential with time. This particular property of constant current source helps for proper signal conditioning with nearly zero reflection from load.

Simple Electronic Circuits

DC Lighting Circuit

A DC supply is used for a small LED that has two terminals namely anode and cathode. The anode is +ve and cathode is −ve. Here, a lamp is used as a load, that has two terminals such as positive and negative. The +ve terminals of the lamp are connected to the anode terminal of the battery and the −ve terminal of the battery is connected to the −ve terminal of the battery. A switch is connected in between wire to give a supply DC voltage to the LED bulb.

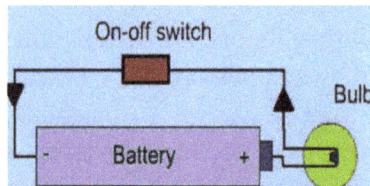

DC Lighting Circuit.

Rain Alarm

The following rain circuit is used to give an alert when it's going to rain. This circuit is used in homes to guard their washed clothes and other things that are vulnerable to rain when they stay in the home most of the time for their work. The required components to build this circuit are probes. 10K and 330K resistors, BC548 and BC 558 transistors, 3V battery, 01mf capacitor and speaker.

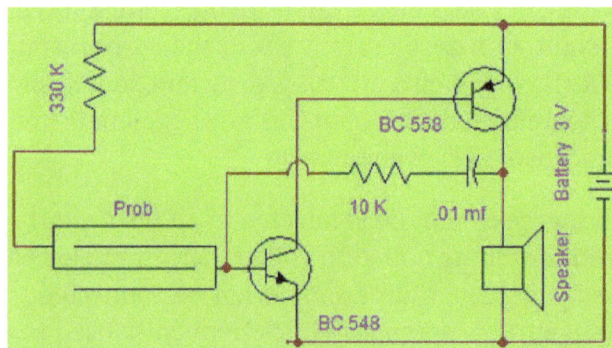

Rain Alarm.

Whenever the rainwater comes in contact with the probe in the above circuit, then the current flows through the circuit to enable the Q1 (NPN) transistor and also Q1 transistor makes Q2 transistor (PNP) to become active. Thus the Q2 transistor conducts and then the flow of current through the speaker generates a buzzer sound. Until the probe is in touch with the water, this procedure replicates again and again. The oscillation circuit built in the above circuit that changes the frequency of the tone, and thus tone can be changed.

Simple Temperature Monitor

This circuit gives an indication using an LED when the battery voltage falls below 9 volts. This circuit is an ideal to monitor the level of charge in 12V small batteries. These batteries are used in burglar alarm systems and portable devices.The working of this circuit depends on the biasing of the base terminal of T1 transistor.

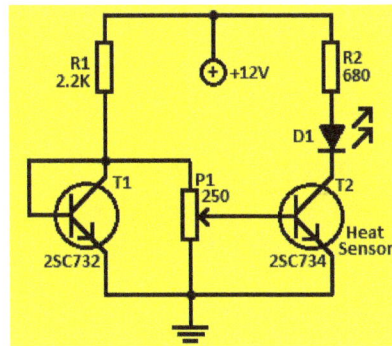

Simple Temperature Monitor.

When the voltage of battery is more than 9 volts, then the voltage on base-emitter terminals will be same. This keeps both transistor and LED off. When the voltage of the battery reduces below 9V due to utilization, the base voltage of T1 transistor falls while its emitter voltage remains same since the C1 capacitor is fully charged.At this stage, base terminal of the T1 transistor becomes +ve and turns ON. C1 capacitor discharges through the LED

Touch Sensor Circuit

The touch sensor circuit is built with three components such as a resistor, a transistor and a light emitting diode. Here, both the resistor and LED connected in series with the positive supply to the collector terminal of the transistor. Select a resistor to set the current of the LED to around 20mA. Now give the connections at the two exposed ends, one connection goes to the +ve supply and another goes to the base terminal of the transistor. Now touch these two wires with your finger. Touch these wires with a finger, then the LED lights up.

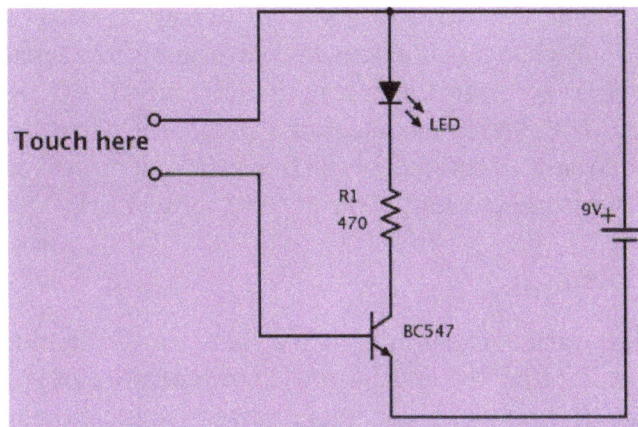

Touch Sensor Circuit.

Multimeter Circuit

A multimeter is a an essential, simple and basic electrical circuit,that is used to measure voltage, resistance and current. It is also used to measure DC as well as AC parameters. Multimeter includes a galvanometer that is connected in series with a resistance. The Voltage across the circuit can be measured by placing the probes of the multimeter across the circuit. The multimeter is mainly used for the continuity of the windings in a motor.

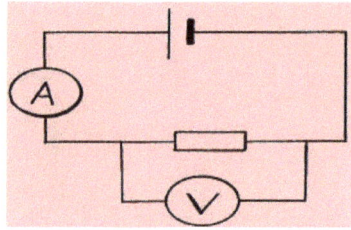

Multimeter Circuit.

LED Flasher Circuit

The circuit configuration of LED flasher is shown below. The following circuit is built with one of the most popular components like the 555 timer and integrated circuits. This circuit will blink the led ON & OFF at regular intervals.

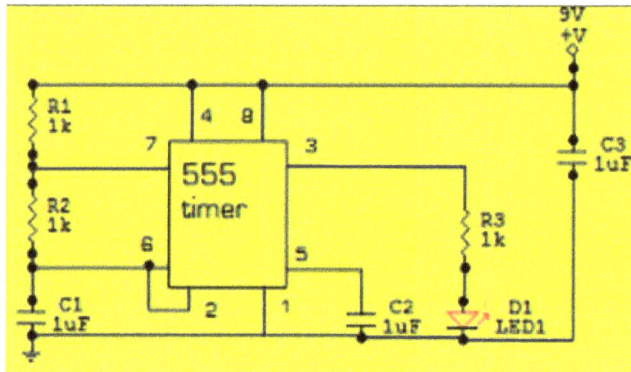

LED Flasher Circuit.

From left to right in the circuit, the capacitor and the two transistors set the time and it takes to switch the LED ON or OFF. By changing the time it takes to charge the capacitor to activate the timer.The IC 555 timer is used to determine the time of the LED stays ON & OFF. It includes a difficult circuit inside, but since it is enclosed in the integrated circuit.The two capacitors are located at the right side of the timer and these are required for the timer to work properly. The last part is the LED and the resistor. The resistor is used to restrict the current on the LED.

Invisible Burglar Alarm

The circuit of the invisible burglar alarm is built with a photo transistor and an IR LED. When there is no obstacle in the path of infrared rays, an alarm will not generate buzzer sound. When somebody crosses the Infrared beam, then an alarm generated buzzer sound. If the photo transistor and the infrared LED are enclosed in black tubes and connected perfectly, the circuit range is 1 meter.

When the infrared beam falls on the L14F1 photo transistor, it performs to keep the BC557 (PNP) out of conduction and the buzzer will not generate the sound in this condition. When the infrared beam breaks, then the photo transistor turns OFF, permitting the PNP transistor to perform and the buzzer sounds. Fix the photo transistor and infrared LED on the reverse sides with correct position to make the buzzer silent. Adjust the variable resistor to set the biasing of the PNP transistor.Here other kinds of photo transistors can also be used instead of LI4F1, but L14F1 is more sensitive.

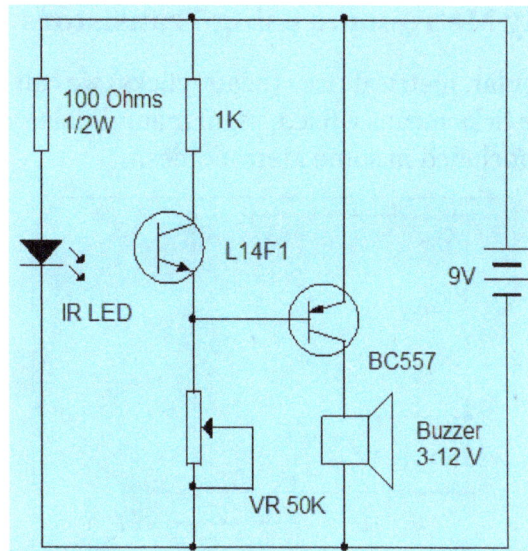

Invisible Burglar Alarm.

LED Circuit

Light Emitting Diode is a small component that gives light. There is a lot of advantages by using LED because it is very cheap, easy to use and we can easily understand whether the circuit is working or not by its indication.

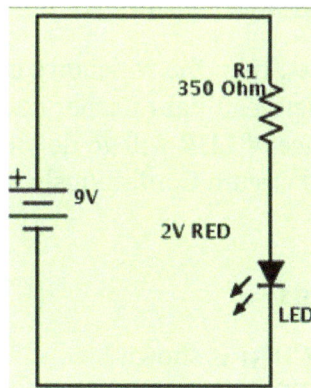

LED Circuit.

Under the forward bias condition, the holes and electrons across the junction move back and forth. In that process, they will get combine or otherwise eliminate one another out. After some time if an electron moves from n-type silicon to p-type silicon, then that electron will get combined with a hole and it will disappear. It makes one complete atom and that is more stable, so it will generate little amount of energy in the form of photons of light.

Under reverse bias condition, the positive power supply will draw away all the electrons present in the junction. And all the holes will draw towards the negative terminal. So the junction is depleted with charge carriers and current will not flow through it.

The anode is the long pin. This is the pin you connect to the most positive voltage. The cathode pin should connect to the most negative voltage. They must be connected correctly for the LED to work.

Simple Light Sensitivity Metronome using Transistors

Any device that produces regular, metrical ticks (beats, clicks) we can call it as Metronome (settable beats per a minute). Here ticks means a fixed, regular aural pulse. Synchronized visual motion like pendulum-swing is also included in some Metronomes.

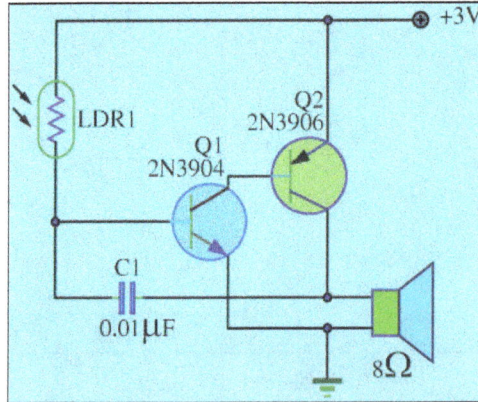

Simple Light Sensitivity Metronome Using Transistors.

This is Simple light sensitivity Metronome circuit using Transistors. Two kinds of transistors are used in this circuit, namely transistor number 2N3904 and 2N3906 make an origin frequency circuit. Sound from a loudspeaker will increase and is down by the frequency in the sound.LDR is used in this circuit LDR means Light Dependent Resistor also we can call it as a photo resistor or photocell. LDR is a light controlled variable resistor.

If the incident light intensity increases, then the resistance of LDR will decrease. This phenomenon is called photo conductivity. When lead light flasher comes to near LDR within a darkroom it receives the light, then the resistance of LDR will go down. That will enhance or affect the frequency of the origin, frequency sound circuit. Continuously wood keeps stroking the music by the frequency change in the circuit.

FM Transmitter using UPC1651

The FM transmitter circuit using UPC1651 is shown below. This circuit is built with UPC1651 IC. This chip is a wide band silicon amplifier, that has a frequency response (1200MHz) and power gain (19dB).

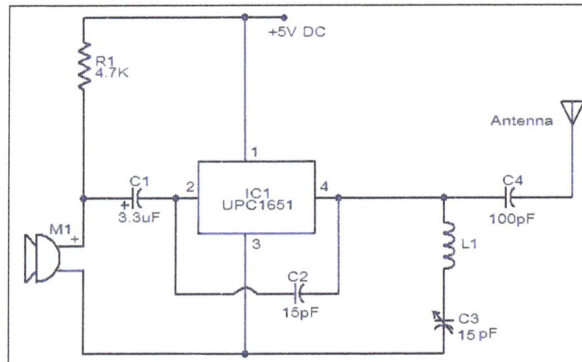

FM Transmitter using UPC1651.

This chip can be worked with 5 volts DC. The received audio signals from the microphone are fed to the i/p pin2 of the chip through the capacitor 'C1'.Here, in the below circuit capacitor acts as a noise filter.

The modulated FM signal will be available at the pin4 (output pin) of the IC. Here, 'C3' capacitor & 'L1' Inductor shapes the required LC circuit for building the oscillations. The transmitter frequency can be altered by regulating the capacitor 'C3'.

References

- What-is-an-electrical-circuit, electricity, science: eschooltoday.com, Retrieved 17 May, 2019

- "Pixii Machine invented by Hippolyte Pixii, National High Magnetic Field Laboratory". Archived from the original on 2008-09-07. Retrieved 2012-03-23

- What-is-an-ac-circuit: circuitglobe.com, Retrieved 19 April, 2019

- Andrew J. Robinson, Lynn Snyder-Mackler (2007). Clinical Electrophysiology: Electrotherapy and Electrophysiologic Testing (3rd ed.). Lippincott Williams & Wilkins. P. 10. ISBN 978-0-7817-4484-3

- What-is-an-electronic-circuit, components, electronics, programming: dummies.com, Retrieved 25 Februray, 2019

- "Pixii Machine invented by Hippolyte Pixii, National High Magnetic Field Laboratory". Archived from the original on 2008-09-07. Retrieved 2008-06-12

- Printed-circuit-board, 2267, definition: techopedia.com, Retrieved 13 May, 2019

- What-are-the-functions-of-a-circuit-board: techwalla.com, Retrieved 8 January, 2019

- Integrated-circuit, entry: newworldencyclopedia.org, Retrieved 5 August, 2019

- Saraju Mohanty, Nanoelectronic Mixed-Signal System Design, mcgraw-Hill, 2015, ISBN 978-0071825719 and 0071825711

- Top-10-simple-electronic-circuits-for-beginners: elprocus.com, Retrieved 14 July, 2019

Circuit: Laws and Theorems

The major laws related to circuits are Kirchhoff's Voltage Law, Kirchhoff's Current Law, Ohm's Law, Joules Law of Heating, etc. The main theorems which are studied in relation to circuits are Thévenin's theorem, Norton's theorem, Reciprocity theorem and Millman's theorem. This chapter has been carefully written to provide an easy understanding of these laws and theorems related to circuits.

KIRCHHOFF'S VOLTAGE LAW

Kirchhoff's Voltage Law (KVL) is Kirchhoff's second law that deals with the conservation of energy around a closed circuit path.

Gustav Kirchhoff's Voltage Law is the second of his fundamental laws we can use for circuit analysis. His voltage law states that for a closed loop series path the algebraic sum of all the voltages around any closed loop in a circuit is equal to zero. This is because a circuit loop is a closed conducting path so no energy is lost.

In other words the algebraic sum of ALL the potential differences around the loop must be equal to zero as: $\Sigma V = 0$. Note here that the term "algebraic sum" means to take into account the polarities and signs of the sources and voltage drops around the loop.

This idea by Kirchhoff is commonly known as the Conservation of Energy, as moving around a closed loop, or circuit, you will end up back to where you started in the circuit and therefore back to the same initial potential with no loss of voltage around the loop. Hence any voltage drops around the loop must be equal to any voltage sources met along the way.

So when applying Kirchhoff's voltage law to a specific circuit element, it is important that we pay special attention to the algebraic signs, (+ and -) of the voltage drops across elements and the emf's of sources otherwise our calculations may be wrong.

A Single Circuit Element

For this simple example assume that the current, I is in the same direction as the flow of positive charge, that is conventional current flow.

Here the flow of current through the resistor is from point A to point B, that is from positive terminal to a negative terminal. Thus as we are travelling in the same direction as current flow, there will be a fall in potential across the resistive element giving rise to a -IR voltage drop across it.

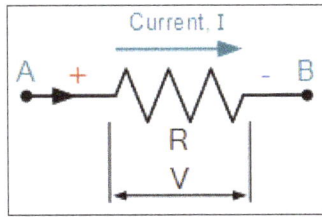

If the flow of current was in the opposite direction from point B to point A, then there would be a rise in potential across the resistive element as we are moving from a - potential to a + potential giving us a +I*R voltage drop.

Thus to apply Kirchhoff's voltage law correctly to a circuit, we must first understand the direction of the polarity and as we can see, the sign of the voltage drop across the resistive element will depend on the direction of the current flowing through it. As a general rule, you will loose potential in the same direction of current across an element and gain potential as you move in the direction of an emf source.

The direction of current flow around a closed circuit can be assumed to be either clockwise or anticlockwise and either one can be chosen. If the direction chosen is different from the actual direction of current flow, the result will still be correct and valid but will result in the algebraic answer having a minus sign.

A Single Circuit Loop

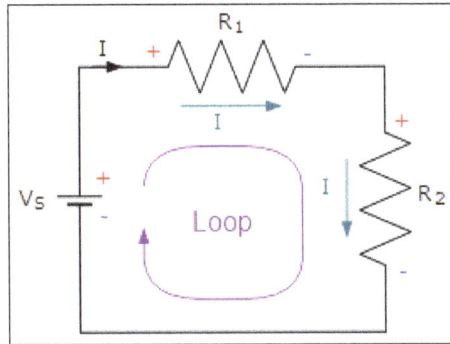

Kirchhoff's voltage law states that the algebraic sum of the potential differences in any loop must be equal to zero as: $\Sigma V = 0$. Since the two resistors, R_1 and R_2 are wired together in a series connection, they are both part of the same loop so the same current must flow through each resistor.

Thus the voltage drop across resistor, $R_1 = I*R_1$ and the voltage drop across resistor, $R_2 = I*R_2$ giving by KVL:

$$V_S + (-IR_1) + (-IR_2) = 0$$
$$\therefore V_S = IR_1 + IR_2$$
$$V_S = I(R_1 + R_2)$$
$$V_S = IR_T$$
$$\text{Where} : R_T = R_1 + R_2$$

We can see that applying Kirchhoff's Voltage Law to this single closed loop produces the formula for the equivalent or total resistance in the series circuit and we can expand on this to find the values of the voltage drops around the loop.

$$R_T = R_1 + R_2$$

$$I = \frac{V_S}{R_T} = \frac{V_S}{R_1 + R_2}$$

$$V_{R1} = IR_1 = V_S\left(\frac{R_1}{R_1 + R_2}\right)$$

$$V_{R2} = IR_2 = V_S\left(\frac{R_2}{R_1 + R_2}\right)$$

Kirchhoff's Voltage Law Example

Three resistor of values: 10 ohms, 20 ohms and 30 ohms, respectively are connected in series across a 12 volt battery supply. Calculate: a) the total resistance, b) the circuit current, c) the current through each resistor, d) the voltage drop across each resistor, e) verify that Kirchhoff's voltage law, KVL holds true.

a) Total Resistance (R_T)

$$R_T = R_1 + R_2 + R_3 = 0\Omega + 20\Omega + 30\Omega = 60\Omega$$

Then the total circuit resistance R_T is equal to 60Ω.

b) Circuit Current (I)

$$I\frac{V_S}{R_T} = \frac{12}{60} = 0.2A$$

Thus the total circuit current I is equal to 0.2 amperes or 200mA.

c) Current Through Each Resistor

The resistors are wired together in series, they are all part of the same loop and therefore each experience the same amount of current. Thus:

$$I_{R1} = I_{R2} = I_{R3} = I_{SERIES} = 0.2 \text{ amperes}$$

d) Voltage Drop Across Each Resistor

$$V_{R1} = I \times R_1 = 0.2 \times 10 = 2 \text{ volts}$$
$$V_{R2} = I \times R_2 = 0.2 \times 20 = 4 \text{ volts}$$
$$V_{R3} = I \times R_3 = 0.2 \times 30 = 6 \text{ volts}$$

e) Verify Kirchhoff's Voltage Law

$$V_S + (-IR_1) + (-IR_2) + (-IR_3) = 0$$
$$12 + (-0.2 \times 10) + (-0.2 \times 20) + (-0.2 \times 30) = 0$$
$$12 + (-2) + (-4) + (-6) = 0$$
$$\therefore 12 - 2 - 4 - 6 = 0$$

Thus Kirchhoff's voltage law holds true as the individual voltage drops around the closed loop add up to the total.

Kirchhoff's Circuit Loop

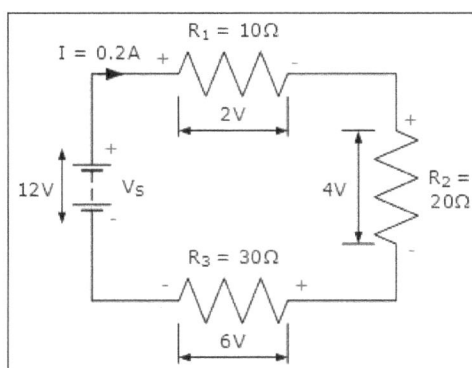

We have seen here that Kirchhoff's voltage law, KVL is Kirchhoff's second law and states that the algebraic sum of all the voltage drops, as you go around a closed circuit from some fixed point and return back to the same point, and taking polarity into account, is always zero. That is $\Sigma V = 0$.

The theory behind Kirchhoff's second law is also known as the law of conservation of voltage, and this is particularly useful for us when dealing with series circuits, as series circuits also act as voltage dividers and the voltage divider circuit is an important application of many series circuits.

KIRCHHOFF'S CURRENT LAW

Gustav Kirchhoff's Current Law is one of the fundamental laws used for circuit analysis. His current law states that for a parallel path the total current entering a circuits junction is exactly equal to the total current leaving the same junction. This is because it has no other place to go as no charge is lost.

In other words the algebraic sum of ALL the currents entering and leaving a junction must be equal to zero as: $\Sigma I_{IN} = \Sigma I_{OUT}$.

This idea by Kirchhoff is commonly known as the Conservation of Charge, as the current is conserved around the junction with no loss of current. Lets look at a simple example of Kirchhoff's current law (KCL) when applied to a single junction.

A Single Junction

Here in this simple single junction example, the current I_T leaving the junction is the algebraic sum of the two currents, I_1 and I_2 entering the same junction. That is $I_T = I_1 + I_2$.

Note that we could also write this correctly as the algebraic sum of: $I_T - (I_1 + I_2) = 0$.

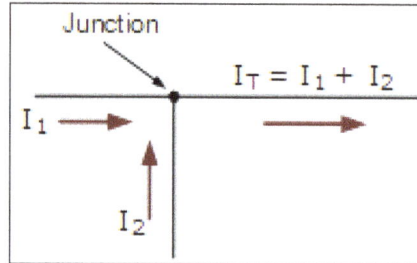

So if I_1 equals 3 amperes and I_2 is equal to 2 amperes, then the total current, I_T leaving the junction will be $3 + 2 = 5$ amperes, and we can use this basic law for any number of junctions or nodes as the sum of the currents both entering and leaving will be the same.

Also, if we reversed the directions of the currents, the resulting equations would still hold true for I_1 or I_2. As $I_1 = I_T - I_2 = 5 - 2 = 3$ amps, and $I_2 = I_T - I_1 = 5 - 3 = 2$ amps. Thus we can think of the currents entering the junction as being positive (+), while the ones leaving the junction as being negative (-).

Then we can see that the mathematical sum of the currents either entering or leaving the junction and in whatever direction will always be equal to zero, and this forms the basis of Kirchhoff's Junction Rule, more commonly known as *Kirchhoff's Current Law*, or (KCL).

Resistors in Parallel

Let's look how we could apply Kirchhoff's current law to resistors in parallel, whether the resistances in those branches are equal or unequal. Consider the following circuit diagram:

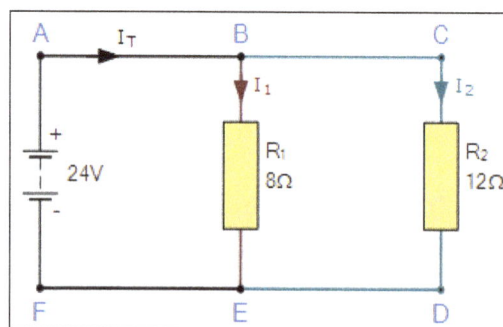

In this simple parallel resistor example there are two distinct junctions for current. Junction one occurs at node B, and junction two occurs at node E. Thus we can use Kirchhoff's Junction Rule for the electrical currents at both of these two distinct junctions, for those currents entering the junction and for those currents flowing leaving the junction.

To start, all the current, I_T leaves the 24 volt supply and arrives at point A and from there it enters node B. Node B is a junction as the current can now split into two distinct directions, with some of the current flowing downwards and through resistor R_1 with the remainder continuing on through resistor R_2 via node C. Note that the currents flowing into and out of a node point are commonly called branch currents.

We can use Ohm's Law to determine the individual branch currents through each resistor as: I = V/R, thus:

For current branch B to E through resistor R_1,

$$I_{B-E} = I_1 = \frac{V}{R_1} = \frac{24}{8} = 3A$$

For current branch C to D through resistor R_2,

$$I_{C-D} = I_2 = \frac{V}{R_2} = \frac{24}{12} = 2A$$

From above we know that Kirchhoff's current law states that the sum of the currents entering a junction must equal the sum of the currents leaving the junction, and in our simple example above, there is one current, I_T going into the junction at node B and two currents leaving the junction, I_1 and I_2.

Since we now know from calculation that the currents leaving the junction at node B is I_1 equals 3 amps and I_2 equals 2 amps, the sum of the currents entering the junction at node B must equal 3 + 2 = 5 amps. Thus $\Sigma_{IN} = I_T = 5$ amperes.

We have two distinct junctions at node B and node E, thus we can confirm this value for I_T as the two currents recombine again at node E. So, for Kirchhoff's junction rule to hold true, the sum of the currents into point F must equal the sum of the currents flowing out of the junction at node E.

As the two currents entering junction E are 3 amps and 2 amps respectively, the sum of the currents entering point F is therefore: 3 + 2 = 5 amperes. Thus $\Sigma_{IN} = I_T = 5$ amperes and therefore Kirchhoff's current law holds true as this is the same value as the current leaving point A.

Applying KCL to more Complex Circuits

We can use Kirchhoff's current law to find the currents flowing around more complex circuits. We hopefully know by now that the algebraic sum of all the currents at a node (junction point) is equal to zero and with this idea in mind, it is a simple case of determining the currents entering a node and those leaving the node. Consider the circuit below.

Kirchhoff's Current Law Example

In this example there are four distinct junctions for current to either separate or merge together at nodes A, C, E and node F. The supply current I_T separates at node A flowing through resistors R_1 and R_2, recombining at node C before separating again through resistors R_3, R_4 and R_5 and finally recombining once again at node F.

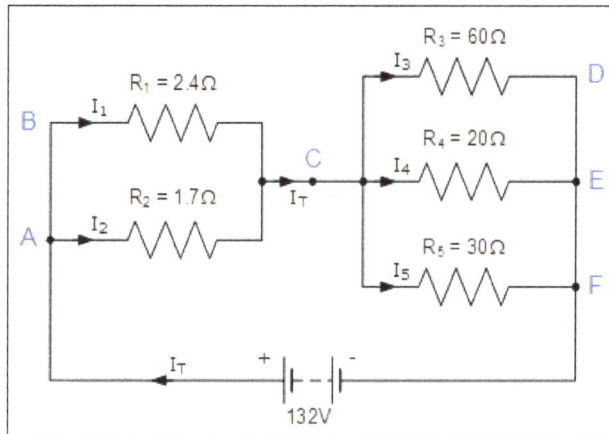

But before we can calculate the individual currents flowing through each resistor branch, we must first calculate the circuits total current, I_T. Ohms law tells us that $I = V/R$ and as we know the value of V, 132 volts, we need to calculate the circuit resistances as follows.

Circuit Resistance R_{AC}

$$\frac{1}{R_{(AC)}} = \frac{1}{R_1} + \frac{1}{R_2} = \frac{1}{2.4} + \frac{1}{1.7}$$

$$\frac{1}{R(AC)} = \frac{1}{1} \qquad \therefore R_{(AC)} = 1\Omega$$

Thus the equivalent circuit resistance between nodes A and C is calculated as 1 Ohm.

Circuit Resistance R_{CF}

$$\frac{1}{R_{(CF)}} = \frac{1}{R_3} + \frac{1}{R_4} + \frac{1}{R_5} = \frac{1}{60} + \frac{1}{20} + \frac{1}{30}$$

$$\frac{1}{R_{(CF)}} = \frac{1}{0.1} \qquad \therefore R_{(CF)} = 10\Omega$$

Thus the equivalent circuit resistance between nodes C and F is calculated as 10 Ohms. Then the total circuit current, I_T is given as:

$$R_T = R_{(AC)} + R_{(CF)} = 1 + 10 = 11\,W$$

$$I_T = \frac{V}{R_T} = \frac{132}{11} = 12\,\text{Amps}$$

Giving us an equivalent circuit of:

Kirchhoff's Current Law Equivalent Circuit

Therefore, V = 132V, R_{AC} = 1Ω, R_{CF} = 10Ω's and I_T = 12A.

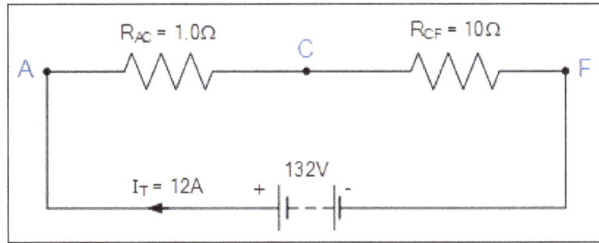

Having established the equivalent parallel resistances and supply current, we can now calculate the individual branch currents and confirm using Kirchhoff's junction rule as follows:

$$V_{AC} = I_T \times R_{AC} = 12 \times 1 = 12 \text{ Volts}$$

$$V_{CF} = I_T \times R_{CF} = 12 \times 10 = 120 \text{ Volts}$$

$$I_1 \frac{V_{AC}}{R_1} = \frac{12}{2.4} = 5 \text{ Amps}$$

$$I_2 = \frac{V_{AC}}{R_2} \cdot \frac{12}{1.7} = 7 \text{ Amps}$$

$$I_3 \frac{V_{CF}}{R_3} = \frac{120}{60} = 2 \text{ Amps}$$

$$I_4 = \frac{V_{CF}}{R_4} = \frac{120}{20} = 6 \text{ Amps}$$

$$I_5 = \frac{V_{CF}}{R_5} = \frac{120}{30} = 4 \text{ Amps}$$

Thus, I_1 = 5A, I_2 = 7A, I_3 = 2A, I_4 = 6A, and I_5 = 4A.

We can confirm that Kirchoff's current law holds true around the circuit by using node C as our reference point to calculate the currents entering and leaving the junction as:

$$\text{At node C } \sum I_{IN} = \sum I_{OUT}$$

$$I_T = I_1 + I_2 = I_3 + I_4 + I_5$$

$$\therefore 12 = (5+7) = (2+6+4)$$

We can also double check to see if Kirchhoffs Current Law holds true as the currents entering the junction are positive, while the ones leaving the junction are negative, thus the algebraic sum is:

$$I_1 + I_2 - I_3 - I_4 - I_5 = 0 \text{ which equals } 5 + 7 - 2 - 6 - 4 = 0.$$

So we can confirm by analysis that Kirchhoff's current law (KCL) which states that the algebraic sum of the currents at a junction point in a circuit network is always zero is true and correct in this example.

Kirchhoff's Current Law Example

Find the currents flowing around the following circuit using Kirchhoff's Current Law only.

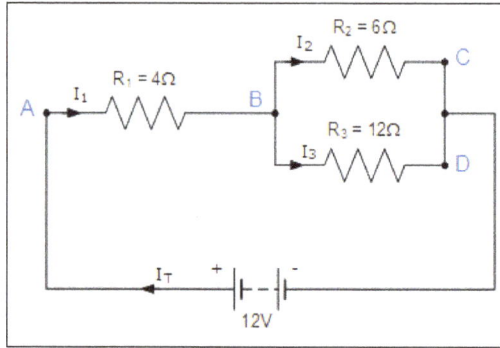

I_T is the total current flowing around the circuit driven by the 12V supply voltage. At point A, I_1 is equal to I_T, thus there will be an I_1*R voltage drop across resistor R_1.

The circuit has 2 branches, 3 nodes (B, C and D) and 2 independent loops, thus the I*R voltage drops around the two loops will be:

- Loop ABC $\Rightarrow 12 = 4I_1 + 6I_2$

- Loop ABD $\Rightarrow 12 = 4I_1 + 12I_3$

Since Kirchhoff's current law states that at node B, $I_1 = I_2 + I_3$, we can therefore substitute current I_1 for ($I_2 + I_3$) in both of the following loop equations and then simplify.

Kirchhoff's Loop Equations

Loop (ABC)   Loop (ABD

$12=4I_1+6I_2$ $12=4I_1+12I_3$
$12=4(I_2+I_3)+6I_2$ $12=4(I_2+I_3)+12I_3$
$12=4I_2+4I_3+6I_2$ $12=4I_2+4I_3+12I_3$
$12=10I_2+4I_3$ $12=4I_2+16I_3$

We now have two simultaneous equations that relate to the currents flowing around the circuit.

$12 = 10I_2 + 4I_3$

$12 = 4I_2 + 16I_3$

By multiplying the first equation (Loop ABC) by 4 and subtracting Loop ABD from Loop ABC, we can be reduced both equations to give us the values of I_2 and I_3,

$12 = 10I_2 + 4I_3 \, (x4) \Rightarrow 48 = 40I_2 + 16I_3$

$12 = 4I_2 + 16I_3 \, (x1) \Rightarrow 12 = 4I_2 + 16I_3$

$48 = 40I_2 + 16I_3 - 12 = 4I_2 + 16I_3 \Rightarrow 36 = 36I_2 + 0$

Substitution of I_2 in terms of I_3 gives us the value of I_2 as 1.0 Amps.

Now we can do the same procedure to find the value of I_3 by multiplying the first equation (Loop ABC) by 4 and the second equation (Loop ABD) by 10. Again by subtracting Loop ABC from Loop ABD, we can be reduced both equations to give us the values of I_2 and I_3,

$$12 = 10I_2 + 4I_3 \ (\text{x}4) \Rightarrow 48 = 40I_2 + 16I_3$$

$$12 = 4I_2 + 16I_3 \ (\text{x}10) \Rightarrow 120 = 40I_2 + 160I_3$$

$$120 = 40I_2 + 160I_3 - 48 = 40I_2 + 16I_3 \Rightarrow 72 = 0 + 144I_3$$

Thus substitution of I_3 in terms of I_2 gives us the value of I_3 as 0.5 Amps.

As Kirchhoff's junction rule states that : $I_1 = I_2 + I_3$

The supply current flowing through resistor R_1 is given as : $1.0 + 0.5 = 1.5$ Amps.

Thus $I_1 = I_T = 1.5$ Amps, $I_2 = 1.0$ Amps and $I_3 = 0.5$ Amps and from that information we could calculate the I*R voltage drops across the devices and at the various points (nodes) around the circuit.

OHM'S LAW

Ohm's law states that the current through a conductor between two points is directly proportional to the voltage across the two points. Introducing the constant of proportionality, the resistance, one arrives at the usual mathematical equation that describes this relationship:

$$I = \frac{V}{R},$$

where I is the current through the conductor in units of amperes, V is the voltage measured *across* the conductor in units of volts, and R is the resistance of the conductor in units of ohms. More specifically, Ohm's law states that the R in this relation is constant, independent of the current. Ohm's law is an empirical relation which accurately describes the conductivity of the vast majority of electrically conductive materials over many orders of magnitude of current. However some materials do not obey Ohm's law, these are called non-ohmic.

The law was named after the German physicist Georg Ohm, who, in a treatise published in 1827, described measurements of applied voltage and current through simple electrical circuits containing various lengths of wire. Ohm explained his experimental results by a slightly more complex equation than the modern form above.

In physics, the term *Ohm's law* is also used to refer to various generalizations of the law; for example the vector form of the law used in electromagnetics and material science:

$$\mathbf{J} = \sigma \mathbf{E},$$

where **J** is the current density at a given location in a resistive material, **E** is the electric field at that location, and σ (sigma) is a material-dependent parameter called the conductivity. This reformulation of Ohm's law is due to Gustav Kirchhoff.

In January 1781, before Georg Ohm's work, Henry Cavendish experimented with Leyden jars and glass tubes of varying diameter and length filled with salt solution. He measured the current by noting how strong a shock he felt as he completed the circuit with his body. Cavendish wrote that the "velocity" (current) varied directly as the "degree of electrification" (voltage). He did not communicate his results to other scientists at the time, and his results were unknown until Maxwell published them in 1879.

Francis Ronalds delineated "intensity" (voltage) and "quantity" (current) for the dry pile – a high voltage source – in 1814 using a gold-leaf electrometer. He found for a dry pile that the relationship between the two parameters was not proportional under certain meteorological conditions.

Ohm did his work on resistance in the years 1825 and 1826, and published his results in 1827 as the book *Die galvanische Kette, mathematisch bearbeitet* ("The galvanic circuit investigated mathematically"). He drew considerable inspiration from Fourier's work on heat conduction in the theoretical explanation of his work. For experiments, he initially used voltaic piles, but later used a thermocouple as this provided a more stable voltage source in terms of internal resistance and constant voltage. He used a galvanometer to measure current, and knew that the voltage between the thermocouple terminals was proportional to the junction temperature. He then added test wires of varying length, diameter, and material to complete the circuit. He found that his data could be modeled through the equation,

$$x = \frac{a}{b+l},$$

where x was the reading from the galvanometer, l was the length of the test conductor, a depended on the thermocouple junction temperature, and b was a constant of the entire setup. From this, Ohm determined his law of proportionality and published his results.

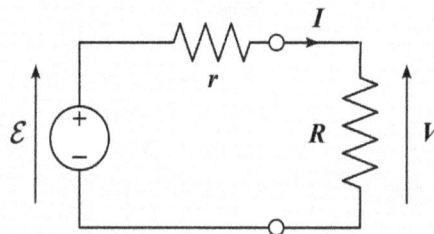

Internal resistance model.

In modern notation we would write,

$$I = \frac{\mathcal{E}}{r+R},$$

where $\mathcal{E}$ is the open-circuit emf of the thermocouple, r is the internal resistance of the thermocouple and R is the resistance of the test wire. In terms of the length of the wire this becomes,

$$I = \frac{\mathcal{E}}{r+\mathcal{R}l},$$

where $\mathcal{R}$ is the resistance of the test wire per unit length. Thus, Ohm's coefficients are,

$$a = \frac{\mathcal{E}}{\mathcal{R}}, \quad b = \frac{r}{\mathcal{R}}.$$

Ohm's law was probably the most important of the early quantitative descriptions of the physics of electricity. We consider it almost obvious today. When Ohm first published his work, this was not the case; critics reacted to his treatment of the subject with hostility. They called his work a "web of naked fancies" and the German Minister of Education proclaimed that "a professor who preached such heresies was unworthy to teach science." The prevailing scientific philosophy in Germany at the time asserted that experiments need not be performed to develop an understanding of nature because nature is so well ordered, and that scientific truths may be deduced through reasoning alone. Also, Ohm's brother Martin, a mathematician, was battling the German educational system. These factors hindered the acceptance of Ohm's work, and his work did not become widely accepted until the 1840s. However, Ohm received recognition for his contributions to science well before he died.

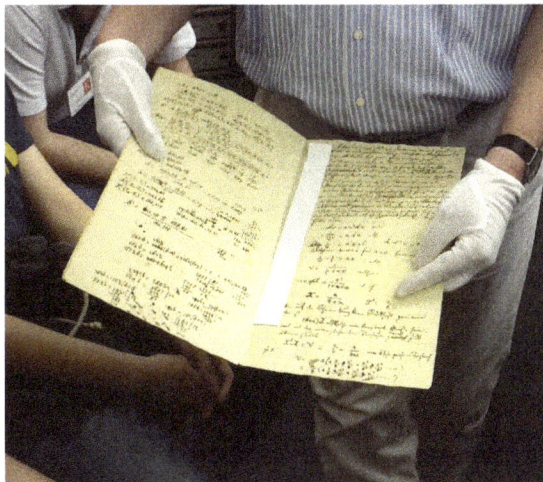

Ohm's law in Georg Ohm's lab book.

In the 1850s, Ohm's law was known as such and was widely considered proved, and alternatives, such as "Barlow's law", were discredited, in terms of real applications to telegraph system design, as discussed by Samuel F. B. Morse in 1855.

The electron was discovered in 1897 by J. J. Thomson, and it was quickly realized that it is the particle (charge carrier) that carries electric currents in electric circuits. In 1900 the first (classical) model of electrical conduction, the Drude model, was proposed by Paul Drude, which finally gave a scientific explanation for Ohm's law. In this model, a solid conductor consists of a stationary lattice of atoms (ions), with conduction electrons moving randomly in it. A voltage across a conductor causes an electric field, which accelerates the electrons in the direction of the electric field, causing a drift of electrons which is the electric current. However the electrons collide with and scatter off of the atoms, which randomizes their motion, thus converting the kinetic energy added to the electron by the field to heat (thermal energy). Using statistical distributions, it can be shown that the average drift velocity of the electrons, and thus the current, is proportional to the electric field, and thus the voltage, over a wide range of voltages.

The development of quantum mechanics in the 1920s modified this picture somewhat, but in modern theories the average drift velocity of electrons can still be shown to be proportional to the electric field, thus deriving Ohm's law. In 1927 Arnold Sommerfeld applied the quantum Fermi-Dirac distribution of electron energies to the Drude model, resulting in the free electron model. A year later, Felix Bloch showed that electrons move in waves (Bloch waves) through a solid crystal lattice, so scattering off the lattice atoms as postulated in the Drude model is not a major process; the electrons scatter off impurity atoms and defects in the material. The final successor, the modern quantum band theory of solids, showed that the electrons in a solid cannot take on any energy as assumed in the Drude model but are restricted to energy bands, with gaps between them of energies that electrons are forbidden to have. The size of the band gap is a characteristic of a particular substance which has a great deal to do with its electrical resistivity, explaining why some substances are electrical conductors, some semiconductors, and some insulators.

While the old term for electrical conductance, the mho (the inverse of the resistance unit ohm), is still used, a new name, the siemens, was adopted in 1971, honoring Ernst Werner von Siemens. The siemens is preferred in formal papers.

In the 1920s, it was discovered that the current through a practical resistor actually has statistical fluctuations, which depend on temperature, even when voltage and resistance are exactly constant; this fluctuation, now known as Johnson–Nyquist noise, is due to the discrete nature of charge. This thermal effect implies that measurements of current and voltage that are taken over sufficiently short periods of time will yield ratios of V/I that fluctuate from the value of R implied by the time average or ensemble average of the measured current; Ohm's law remains correct for the average current, in the case of ordinary resistive materials.

Ohm's work long preceded Maxwell's equations and any understanding of frequency-dependent effects in AC circuits. Modern developments in electromagnetic theory and circuit theory do not contradict Ohm's law when they are evaluated within the appropriate limits.

Scope

Ohm's law is an empirical law, a generalization from many experiments that have shown that current is approximately proportional to electric field for most materials. It is less fundamental than Maxwell's equations and is not always obeyed. Any given material will break down under a strong-enough electric field, and some materials of interest in electrical engineering are "non-ohmic" under weak fields.

Ohm's law has been observed on a wide range of length scales. In the early 20th century, it was thought that Ohm's law would fail at the atomic scale, but experiments have not borne out this expectation. As of 2012, researchers have demonstrated that Ohm's law works for silicon wires as small as four atoms wide and one atom high.

Microscopic Origins

The dependence of the current density on the applied electric field is essentially quantum mechanical in nature; A qualitative description leading to Ohm's law can be based upon classical mechanics using the Drude model developed by Paul Drude in 1900.

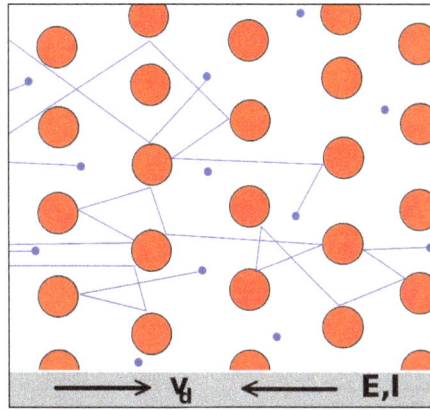

Drude Model electrons (shown here in blue) constantly bounce
among heavier, stationary crystal ions (shown in red).

The Drude model treats electrons (or other charge carriers) like pinballs bouncing among the ions
that make up the structure of the material. Electrons will be accelerated in the opposite direction
to the electric field by the average electric field at their location. With each collision, though, the
electron is deflected in a random direction with a velocity that is much larger than the velocity
gained by the electric field. The net result is that electrons take a zigzag path due to the collisions,
but generally drift in a direction opposing the electric field.

The drift velocity then determines the electric current density and its relationship to E and is
independent of the collisions. Drude calculated the average drift velocity from $p = -eE\tau$ where
p is the average momentum, $-e$ is the charge of the electron and τ is the average time between
the collisions. Since both the momentum and the current density are proportional to the drift
velocity, the current density becomes proportional to the applied electric field; this leads to
Ohm's law.

Hydraulic Analogy

A hydraulic analogy is sometimes used to describe Ohm's law. Water pressure, measured by pas-
cals (or PSI), is the analog of voltage because establishing a water pressure difference between
two points along a (horizontal) pipe causes water to flow. Water flow rate, as in liters per second,
is the analog of current, as in coulombs per second. Finally, flow restrictors—such as apertures
placed in pipes between points where the water pressure is measured—are the analog of resistors.
We say that the rate of water flow through an aperture restrictor is proportional to the difference
in water pressure across the restrictor. Similarly, the rate of flow of electrical charge, that is, the
electric current, through an electrical resistor is proportional to the difference in voltage measured across the resistor.

Flow and pressure variables can be calculated in fluid flow network with the use of the hydraulic
ohm analogy. The method can be applied to both steady and transient flow situations. In the linear
laminar flow region, Poiseuille's law describes the hydraulic resistance of a pipe, but in the turbu-
lent flow region the pressure–flow relations become nonlinear.

The hydraulic analogy to Ohm's law has been used, for example, to approximate blood flow through
the circulatory system.

Circuit Analysis

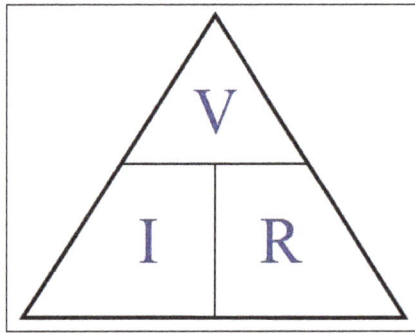

Ohm's law triangle

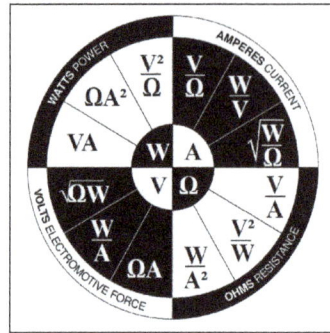

Ohm's law wheel with international unit symbols

In circuit analysis, three equivalent expressions of Ohm's law are used interchangeably:

$$I = \frac{V}{R} \quad \text{or} \quad V = IR \quad \text{or} \quad R = \frac{V}{I}.$$

Each equation is quoted by some sources as the defining relationship of Ohm's law, or all three are quoted, or derived from a proportional form, or even just the two that do not correspond to Ohm's original statement may sometimes be given.

The interchangeability of the equation may be represented by a triangle, where V (voltage) is placed on the top section, the I (current) is placed to the left section, and the R (resistance) is placed to the right. The line that divides the left and right sections indicates multiplication, and the divider between the top and bottom sections indicates division (hence the division bar).

Resistive Circuits

Resistors are circuit elements that impede the passage of electric charge in agreement with Ohm's law, and are designed to have a specific resistance value R. In a schematic diagram the resistor is shown as a zig-zag symbol. An element (resistor or conductor) that behaves according to Ohm's law over some operating range is referred to as an *ohmic device* (or an *ohmic resistor*) because Ohm's law and a single value for the resistance suffice to describe the behavior of the device over that range.

Ohm's law holds for circuits containing only resistive elements (no capacitances or inductances) for all forms of driving voltage or current, regardless of whether the driving voltage or current is constant (DC) or time-varying such as AC. At any instant of time Ohm's law is valid for such circuits.

Resistors which are in *series* or in *parallel* may be grouped together into a single "equivalent resistance" in order to apply Ohm's law in analyzing the circuit.

Reactive Circuits with Time-varying Signals

When reactive elements such as capacitors, inductors, or transmission lines are involved in a circuit to which AC or time-varying voltage or current is applied, the relationship between voltage and current becomes the solution to a differential equation, so Ohm's law does not directly apply since that form contains only resistances having value R, not complex impedances which may contain capacitance ("C") or inductance ("L").

Equations for time-invariant AC circuits take the same form as Ohm's law. However, the variables are generalized to complex numbers and the current and voltage waveforms are complex exponentials.

In this approach, a voltage or current waveform takes the form Ae^{st}, where t is time, s is a complex parameter, and A is a complex scalar. In any linear time-invariant system, all of the currents and voltages can be expressed with the same s parameter as the input to the system, allowing the time-varying complex exponential term to be canceled out and the system described algebraically in terms of the complex scalars in the current and voltage waveforms.

The complex generalization of resistance is impedance, usually denoted Z; it can be shown that for an inductor,

$$Z = sL$$

and for a capacitor,

$$Z = \frac{1}{sC}.$$

We can now write,

$$V = I \cdot Z$$

where V and I are the complex scalars in the voltage and current respectively and Z is the complex impedance.

This form of Ohm's law, with Z taking the place of R, generalizes the simpler form. When Z is complex, only the real part is responsible for dissipating heat.

In the general AC circuit, Z varies strongly with the frequency parameter s, and so also will the relationship between voltage and current.

For the common case of a steady sinusoid, the s parameter is taken to be $j\omega$, corresponding to a complex sinusoid $Ae^{j\omega t}$. The real parts of such complex current and voltage waveforms describe the actual sinusoidal currents and voltages in a circuit, which can be in different phases due to the different complex scalars.

Linear Approximations

Ohm's law is one of the basic equations used in the analysis of electrical circuits. It applies to both metal conductors and circuit components (resistors) specifically made for this behaviour. Both are ubiquitous in electrical engineering. Materials and components that obey Ohm's law are described as "ohmic" which means they produce the same value for resistance (R = V/I) regardless of the value of V or I which is applied and whether the applied voltage or current is DC (direct current) of either positive or negative polarity or AC (alternating current).

In a true ohmic device, the same value of resistance will be calculated from R = V/I regardless of the value of the applied voltage V. That is, the ratio of V/I is constant, and when current is plotted as a function of voltage the curve is *linear* (a straight line). If voltage is forced to some value V, then that voltage V divided by measured current I will equal R. Or if the current is forced to some value I, then the measured voltage V divided by that current I is also R. Since the plot of I versus V is a straight line, then it is also true that for any set of two different voltages V_1 and V_2 applied across a given device of resistance R, producing currents $I_1 = V_1/R$ and $I_2 = V_2/R$, that the ratio $(V_1-V_2)/(I_1-I_2)$ is also a constant equal to R. The operator "delta" (Δ) is used to represent a difference in a quantity, so we can write $\Delta V = V_1-V_2$ and $\Delta I = I_1-I_2$. Summarizing, for any truly ohmic device having resistance R, $V/I = \Delta V/\Delta I = R$ for any applied voltage or current or for the difference between any set of applied voltages or currents.

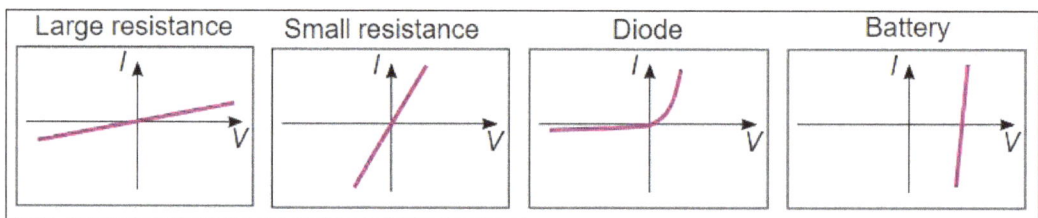

The I–V curves of four devices: Two resistors, a diode, and a battery. The two resistors follow Ohm's law: The plot is a straight line through the origin. The other two devices do *not* follow Ohm's law.

There are, however, components of electrical circuits which do not obey Ohm's law; that is, their relationship between current and voltage (their I–V curve) is *nonlinear* (or non-ohmic). An example is the p-n junction diode (curve at right). As seen in the figure, the current does not increase linearly with applied voltage for a diode. One can determine a value of current (I) for a given value of applied voltage (V) from the curve, but not from Ohm's law, since the value of "resistance" is not constant as a function of applied voltage. Further, the current only increases significantly if the applied voltage is positive, not negative. The ratio *V/I* for some point along the nonlinear curve is sometimes called the *static*, or *chordal*, or DC, resistance, but as seen in the figure the value of total *V* over total *I* varies depending on the particular point along the nonlinear curve which is chosen. This means the "DC resistance" V/I at some point on the curve is not the same as what would be determined by applying an AC signal having peak amplitude ΔV volts or ΔI amps centered at that same point along the curve and measuring $\Delta V/\Delta I$. However, in some diode applications, the AC signal applied to the device is small and it is possible to analyze the circuit in terms of the *dynamic, small-signal,* or *incremental* resistance, defined as the one over the slope of the V–I curve at the average value (DC operating point) of the voltage (that is, one over the derivative of current with respect to voltage). For sufficiently small signals, the dynamic resistance allows the Ohm's law small signal resistance to be calculated as approximately one over the slope of a line drawn tangentially to the V-I curve at the DC operating point.

Temperature Effects

Ohm's law has sometimes been stated as, "for a conductor in a given state, the electromotive force is proportional to the current produced." That is, that the resistance, the ratio of the applied electromotive force (or voltage) to the current, "does not vary with the current strength." The qualifier "in a given state" is usually interpreted as meaning "at a constant temperature," since the resistivity of materials is usually temperature dependent. Because the conduction of current is related to Joule heating of the conducting body, according to Joule's first law, the temperature of a conducting body may change when it carries a current. The dependence of resistance on temperature therefore makes resistance depend upon the current in a typical experimental setup, making the law in this form difficult to directly verify. Maxwell and others worked out several methods to test the law experimentally in 1876, controlling for heating effects.

Relation to Heat Conductions

Ohm's principle predicts the flow of electrical charge (i.e. current) in electrical conductors when subjected to the influence of voltage differences; Jean-Baptiste-Joseph Fourier's principle predicts the flow of heat in heat conductors when subjected to the influence of temperature differences.

The same equation describes both phenomena, the equation's variables taking on different meanings in the two cases. Specifically, solving a heat conduction (Fourier) problem with *temperature* (the driving "force") and *flux of heat* (the rate of flow of the driven "quantity", i.e. heat energy) variables also solves an analogous electrical conduction (Ohm) problem having *electric potential* (the driving "force") and *electric current* (the rate of flow of the driven "quantity", i.e. charge) variables.

The basis of Fourier's work was his clear conception and definition of thermal conductivity. He assumed that, all else being the same, the flux of heat is strictly proportional to the gradient of temperature. Although undoubtedly true for small temperature gradients, strictly proportional behavior will be lost when real materials (e.g. ones having a thermal conductivity that is a function of temperature) are subjected to large temperature gradients.

A similar assumption is made in the statement of Ohm's law: other things being alike, the strength of the current at each point is proportional to the gradient of electric potential. The accuracy of the assumption that flow is proportional to the gradient is more readily tested, using modern measurement methods, for the electrical case than for the heat case.

Other Versions

Ohm's law, in the form above, is an extremely useful equation in the field of electrical/electronic engineering because it describes how voltage, current and resistance are interrelated on a "macroscopic" level, that is, commonly, as circuit elements in an electrical circuit. Physicists who study the electrical properties of matter at the microscopic level use a closely related and more general vector equation, sometimes also referred to as Ohm's law, having variables that are closely related to the V, I, and R scalar variables of Ohm's law, but which are each functions of position within the conductor. Physicists often use this continuum form of Ohm's Law:

$$\mathbf{E} = \rho\mathbf{J}$$

where "E" is the electric field vector with units of volts per meter (analogous to "V" of Ohm's law which has units of volts), "J" is the current density vector with units of amperes per unit area (analogous to "I" of Ohm's law which has units of amperes), and "ρ" is the resistivity with units of ohm·meters (analogous to "R" of Ohm's law which has units of ohms). The above equation is sometimes written as J = E where "σ" is the conductivity which is the reciprocal of ρ.

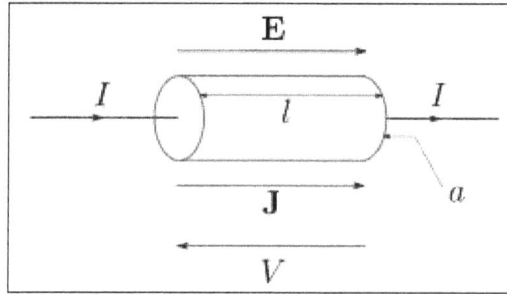

Current flowing through a uniform cylindrical conductor (such as a round wire) with a uniform field applied.

The voltage between two points is defined as:

$$\Delta V = -\int \mathbf{E} \cdot d\mathbf{l}$$

with $d\mathbf{l}$ the element of path along the integration of electric field vector **E**. If the applied **E** field is uniform and oriented along the length of the conductor as shown in the figure, then defining the voltage V in the usual convention of being opposite in direction to the field, and with the understanding that the voltage V is measured differentially across the length of the conductor allowing us to drop the Δ symbol, the above vector equation reduces to the scalar equation:

$$V = El \ \text{ or } \ E = \frac{V}{l}.$$

Since the **E** field is uniform in the direction of wire length, for a conductor having uniformly consistent resistivity ρ, the current density **J** will also be uniform in any cross-sectional area and oriented in the direction of wire length, so we may write:

$$J = \frac{I}{a}.$$

Substituting the above 2 results (for E and J respectively) into the continuum form:

$$\frac{V}{l} = \frac{I}{a} \rho \quad \text{or} \quad V = I \rho \frac{l}{a}.$$

The electrical resistance of a uniform conductor is given in terms of resistivity by:

$$R = \rho \frac{l}{a}$$

where l is the length of the conductor in SI units of meters, a is the cross-sectional area (for a round wire $a = \pi r^2$ if r is radius) in units of meters squared, and ρ is the resistivity in units of ohm·meters.

After substitution of R from the above equation into the equation preceding it, the continuum form of Ohm's law for a uniform field (and uniform current density) oriented along the length of the conductor reduces to the more familiar form:

$$V = IR.$$

A perfect crystal lattice, with low enough thermal motion and no deviations from periodic structure, would have no resistivity, but a real metal has crystallographic defects, impurities, multiple isotopes, and thermal motion of the atoms. Electrons scatter from all of these, resulting in resistance to their flow.

The more complex generalized forms of Ohm's law are important to condensed matter physics, which studies the properties of matter and, in particular, its electronic structure. In broad terms, they fall under the topic of constitutive equations and the theory of transport coefficients.

Magnetic Effects

If an external B-field is present and the conductor is not at rest but moving at velocity $\mathbf{v}$, then an extra term must be added to account for the current induced by the Lorentz force on the charge carriers.

$$\mathbf{J} = \sigma(\mathbf{E} + \mathbf{v} \times \mathbf{B})$$

In the rest frame of the moving conductor this term drops out because $\mathbf{v} = 0$. There is no contradiction because the electric field in the rest frame differs from the E-field in the lab frame: $E' = E + v \times B$. Electric and magnetic fields are relative.

If the current J is alternating because the applied voltage or E-field varies in time, then reactance must be added to resistance to account for self-inductance. The reactance may be strong if the frequency is high or the conductor is coiled.

Conductive Fluids

In a conductive fluid, such as a plasma, there is a similar effect. Consider a fluid moving with the velocity v in a magnetic field B. The relative motion induces an electric field E which exerts electric force on the charged particles giving rise to an electric current J. The equation of motion for the electron gas, with a number density n_e, is written as,

$$m_e n_e \frac{d\mathbf{v}_e}{dt} = -n_e e\mathbf{E} + n_e m_e \nu (v_i - v_e) - e n_e \mathbf{v}_e \times \mathbf{B},$$

where e, m_e and $\mathbf{v}_e$ are the charge, mass and velocity of the electrons, respectively. Also, ν is the frequency of collisions of the electrons with ions which have a velocity field $\mathbf{v}_i$. Since, the electron has a very small mass compared with that of ions, we can ignore the left hand side of the above equation to write,

$$\sigma(\mathbf{E} + \mathbf{v} \times \mathbf{B}) = \mathbf{J},$$

where we have used the definition of the current density, and also put $\sigma = \dfrac{n_e e^2}{\nu m_e}$ which is the electrical conductivity. This equation can also be equivalently written as,

$$\mathbf{E} + \mathbf{v} \times \mathbf{B} = \rho \mathbf{J},$$

where $\rho = \sigma^{-1}$ is the electrical resistivity. It is also common to write η instead of ρ which can be confusing since it is the same notation used for the magnetic diffusivity defined as $\eta = 1/\mu_0\sigma$.

JOULES LAW OF HEATING

When current flows through an electric circuit, the collision between the electrons and atoms of wire causes heat to be generated. How much heat is generated during current flowing through a wire and on what conditions and parameters do the heat generation depend? James Prescott Joule an English physicist, coined a formula which explains this phenomenon accurately. This is known as Joule's law.

The heat which is produced due to the flow of current within an electric wire, is expressed in unit of Joules. Now the mathematical representation and explanation of Joule's law is given in the following manner.

1. The amount of heat produced in a current conducting wire, is proportional to the square of the amount of current that is flowing through the wire, when the electrical resistance of the wire and the time of current flowing are constant.

   i.e. $H \propto i^2$ [When R and t are costant]

2. The amount of heat produced is proportional to the electrical resistance of the wire when the current in the wire and the time of current flowing are constant.

   i.e. $H \propto R$ [When i and t are costant]

3. The heat generated due to the flow of current is proportional to the time of current flowing, when the electrical resistance and the amount of current is constant.

   i.e. $H \propto t$ [When i and R are costant]

When these three conditions are merged, the resulting formula is like this:

$H \propto i^2.R.t$ [When i, R and t all variable]

$H = \dfrac{1}{J}.i^2.R.t$ [$\because$ J is a joule contant]

Here, 'H' is the heat generated in Joules, 'i' is the current flowing through the conducting wire in ampere and 't' is the time in seconds. There are four variables in the equation. When any three of these are known the other one can be calculated. Here, 'J' is a constant, known as Joule's mechanical equivalent of heat. Mechanical equivalent of heat may be defined as the number of work units which, when completely converted into heat, furnishes one unit of heat. Obviously, the value of J will depend on the choice of units of work and heat. It has been found that J = 4.2 joules/cal (1 joule = 10^7 ergs) = 1400 ft. lbs./CHU = 778 ft. lbs/B Th U. It should be noted that the above values are not very accurate but are good enough for general work.

$$There\ fore, H\frac{I^2Rt\ Joules}{4.2\ joules\,/\,cal}=\frac{I^2Rt}{4.2}cal=0.24I^2Rt\ cal$$

Now according to Joule's law I²Rt = work done in joules electrically when I ampere of current are maintained through a resistor of R ohms for t seconds.

$$H=0.24V\ It\ cal=0.24\frac{V^2}{R}cal(as\ V=IR)$$

By eliminating I and R in turn in the above expression with the help of Ohm's law, we get alternative forms as.

THÉVENIN'S THEOREM

As originally stated in terms of DC resistive circuits only, Thévenin's theorem (aka Helmholtz–Thévenin theorem) holds that:

- Any linear electrical network containing only voltage sources, current sources and resistances can be replaced at terminals A-B by an equivalent combination of a voltage source V_{th} in a series connection with a resistance R_{th}.

- The equivalent voltage V_{th} is the voltage obtained at terminals A-B of the network with terminals A-B open circuited.

- The equivalent resistance R_{th} is the resistance that the circuit between terminals A and B would have if all ideal voltage sources in the circuit were replaced by a short circuit and all ideal current sources were replaced by an open circuit.

- If terminals A and B are connected to one another, the current flowing from A to B will be V_{th}/R_{th}. This means that R_{th} could alternatively be calculated as V_{th} divided by the short-circuit current between A and B when they are connected together.

In circuit theory terms, the theorem allows any one-port network to be reduced to a single voltage source and a single impedance.

The theorem also applies to frequency domain AC circuits consisting of reactive and resistive impedances. It means the theorem applies for AC in an exactly same way to DC except that resistances are generalized to impedances.

The theorem was independently derived in 1853 by the German scientist Hermann von Helmholtz and in 1883 by Léon Charles Thévenin (1857–1926), an electrical engineer with France's national Postes et Télégraphes telecommunications organization.

Thévenin's theorem and its dual, Norton's theorem, are widely used to make circuit analysis simpler and to study a circuit's initial-condition and steady-state response. Thévenin's theorem can be used to convert any circuit's sources and impedances to a Thévenin equivalent; use of the theorem may in some cases be more convenient than use of Kirchhoff's circuit laws.

Calculating the Thévenin Equivalent

The equivalent circuit is a voltage source with voltage V_{Th} in series with a resistance R_{Th}.

The Thévenin-equivalent voltage V_{Th} is the open-circuit voltage at the output terminals of the original circuit. When calculating a Thévenin-equivalent voltage, the voltage divider principle is often useful, by declaring one terminal to be V_{out} and the other terminal to be at the ground point.

The Thévenin-equivalent resistance R_{Th} is the resistance measured across points A and B "looking back" into the circuit. The resistance is measured after replacing all voltage- and current-sources with their internal resistances. That means an ideal voltage source is replaced with a short circuit, and an ideal current source is replaced with an open circuit. Resistance can then be calculated across the terminals using the formulae for series and parallel circuits. This method is valid only for circuits with independent sources. If there are dependent sources in the circuit, another method must be used such as connecting a test source across A and B and calculating the voltage across or current through the test source.

The replacements of voltage and current sources do what the sources would do if their values were set to zero. A zero valued voltage source would create a potential difference of zero volts between its terminals, regardless of the current that passes through it; its replacement, a short circuit, does the same thing. A zero valued current source passes zero current, regardless of the voltage across it; its replacement, an open circuit, does the same thing.

Example:

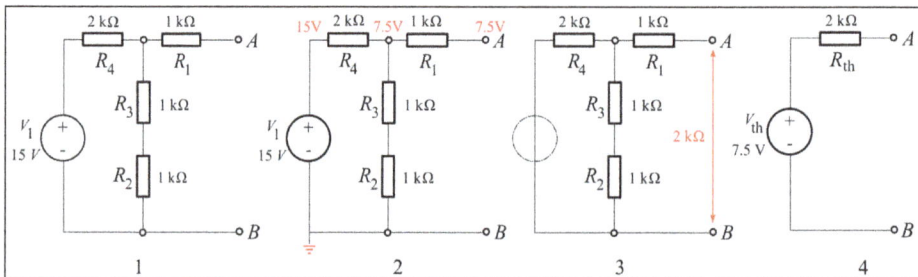

1. Original circuit.

2. The equivalent voltage.

3. The equivalent resistance.

4. The equivalent circuit.

In the example, calculating the equivalent voltage:

$$V_{Th} = \frac{R_2 + R_3}{(R_2 + R_3) + R_4} \cdot V_1$$

$$= \frac{1k\Omega + 1k\Omega}{(1k\Omega + 1k\Omega) + 2k\Omega} \cdot 15V$$

$$= \frac{1}{2} \cdot 15V = 7.5V$$

(notice that R_1 is not taken into consideration, as above calculations are done in an open-circuit condition between A and B, therefore no current flows through this part, which means there is no current through R_1 and therefore no voltage drop along this part).

Calculating equivalent resistance ($R_x \| R_y$ is the total resistance of two parallel resistors):

$$R_{Th} = R_1 + \left[(R_2 + R_3) \| R_4 \right]$$

$$= 1k\Omega + \left[(1k\Omega + 1k\Omega) \| 2k\Omega \right]$$

$$= 1k\Omega + \left(\frac{1}{(1k\Omega + 1k\Omega)} + \frac{1}{(2k\Omega)} \right)^{-1} = 2k\Omega$$

Conversion to a Norton Equivalent

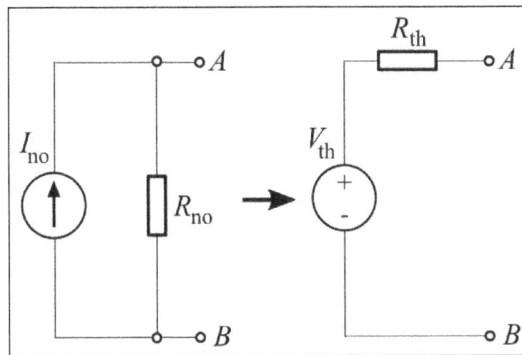

Norton-Thevenin conversion.

A Norton equivalent circuit is related to the Thévenin equivalent by,

$$R_{Th} = R_{No}$$
$$V_{Th} = I_{No} R_{No}$$
$$I_{No} = V_{Th} / R_{Th}$$

Practical Limitations

- Many circuits are only linear over a certain range of values, thus the Thévenin equivalent is valid only within this linear range.

- The Thévenin equivalent has an equivalent I–V characteristic only from the point of view of the load.

- The power dissipation of the Thévenin equivalent is not necessarily identical to the power dissipation of the real system. However, the power dissipated by an external resistor between the two output terminals is the same regardless of how the internal circuit is implemented.

A Proof of the Theorem

The proof involves two steps. The first step is to use superposition theorem to construct a solution. Then, uniqueness theorem is employed to show that the obtained solution is unique. It is noted that the second step is usually implied in literature.

By using superposition of specific configurations, it can be shown that for any linear "black box" circuit which contains voltage sources and resistors, its voltage is a linear function of the corresponding current as follows:

$$V = V_{Eq} - Z_{Eq} I.$$

Here, the first term reflects the linear summation of contributions from each voltage source, while the second term measures the contributions from all the resistors. The above expression is obtained by using the fact that the voltage of the black box for a given current I is identical to the linear superposition of the solutions of the following problems: (1) to leave the black box open circuited but activate individual voltage source one at a time and, (2) to short circuit all the voltage sources but feed the circuit with a certain ideal voltage source so that the resulting current exactly reads I (Alternatively, one can use an ideal current source of current). Moreover, it is straightforward to show that V_{Eq} and Z_{Eq} are the single voltage source and the single series resistor in question.

As a matter of fact, the above relation between V and I is established by superposition of some particular configurations. Now, the uniqueness theorem guarantees that the result is general. To be specific, there is one and only one value of   once the value of I is given. In other words, the above relation holds true independent of what the "black box" is plugged to.

NORTON'S THEOREM

Norton's theorem (aka Mayer–Norton theorem) holds, to illustrate in DC circuit theory terms:

- Any linear electrical network with voltage and current sources and only resistances can be replaced at terminals A–B by an equivalent current source I_{no} in parallel connection with an equivalent resistance R_{no}.

- This equivalent current I_{no} is the current obtained at terminals A-B of the network with terminals A-B short circuited.

- This equivalent resistance R_{no} is the resistance obtained at terminals A-B of the network with all its voltage sources short circuited and all its current sources open circuited.

For alternating current (AC) systems the theorem can be applied to reactive impedances as well as resistances.

The Norton equivalent circuit is used to represent any network of linear sources and impedances at a given frequency.

Norton's theorem and its dual, Thévenin's theorem, are widely used for circuit analysis simplification and to study circuit's initial-condition and steady-state response.

Norton's theorem was independently derived in 1926 by Siemens & Halske researcher Hans Ferdinand Mayer and Bell Labs engineer Edward Lawry Norton.

To find the equivalent,

1. Find the Norton current I_{no}. Calculate the output current, I_{AB}, with a short circuit as the load (meaning 0 resistance between A and B). This is I_{no}.

2. Find the Norton resistance R_{no}. When there are no dependent sources (all current and voltage sources are independent), there are two methods of determining the Norton impedance R_{no}.

   - Calculate the output voltage, V_{AB}, when in open circuit condition (i.e., no load resistor – meaning infinite load resistance). R_{no} equals this V_{AB} divided by I_{no}.

   or

   - Replace independent voltage sources with short circuits and independent current sources with open circuits. The total resistance across the output port is the Norton impedance R_{no}.

This is equivalent to calculating the Thevenin resistance.

However, when there are dependent sources, the more general method must be used. This method is not shown below in the diagrams.

   - Connect a constant current source at the output terminals of the circuit with a value of 1 ampere and calculate the voltage at its terminals. This voltage divided by the 1 A current is the Norton impedance R_{no}. This method must be used if the circuit contains dependent sources, but it can be used in all cases even when there are no dependent sources.

Example of a Norton Equivalent Circuit

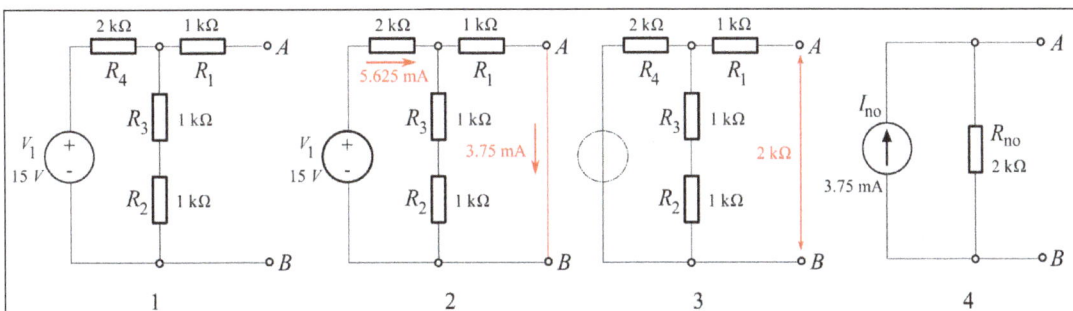

1. The original circuit.

2. Calculating the equivalent output current.

3. Calculating the equivalent resistance.

4. Design the Norton equivalent circuit.

In the example, the total current I_{total} is given by:

$$I_{total} = \frac{15\text{V}}{2\text{k}\Omega + 1\text{k}\Omega \| (1\text{k}\Omega + 1\text{k}\Omega)} = 5.625\text{mA}.$$

The current through the load is then, using the current divider rule:

$$I_{no} = \frac{1\text{k}\Omega + 1\text{k}\Omega}{(1\text{k}\Omega + 1\text{k}\Omega + 1\text{k}\Omega)} \cdot I_{total}$$

$$= 2/3 \cdot 5.625\text{mA} = 3.75\text{mA}.$$

and the equivalent resistance looking back into the circuit is:

$$R_{no} = 1\text{k}\Omega + (2\text{k}\Omega \| (1\text{k}\Omega + 1\text{k}\Omega)) = 2\text{k}\Omega.$$

So the equivalent circuit is a 3.75 mA current source in parallel with a 2 kΩ resistor.

Conversion to a Thévenin Equivalent

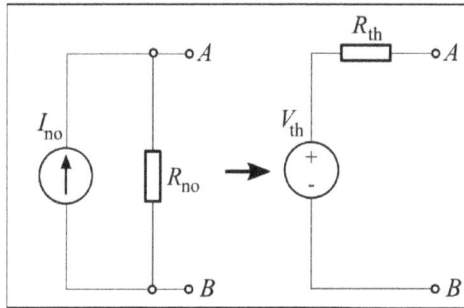

To a Thévenin equivalent.

A Norton equivalent circuit is related to the Thévenin equivalent by the equations:

$$R_{th} = R_{no}$$
$$V_{th} = I_{no} R_{no}$$
$$\frac{V_{th}}{R_{th}} = I_{no}$$

Queueing Theory

The passive circuit equivalent of "Norton's theorem" in queuing theory is called the Chandy Herzog

Woo theorem. In a reversible queueing system, it is often possible to replace an uninteresting subset of queues by a single (FCFS or PS) queue with an appropriately chosen service rate.

SUPER POSITION THEOREMS

The superposition theorem for electrical circuits states that for a linear system the response (voltage or current) in any branch of a bilateral linear circuit having more than one independent source equals the algebraic sum of the responses caused by each independent source acting alone, where all the other independent sources are replaced by their internal impedances.

To ascertain the contribution of each individual source, all of the other sources first must be "turned off" (set to zero) by:

- Replacing all other independent voltage sources with a short circuit (thereby eliminating difference of potential i.e. $V=0$; internal impedance of ideal voltage source is zero (short circuit)).

- Replacing all other independent current sources with an open circuit (thereby eliminating current i.e. $I=0$; internal impedance of ideal current source is infinite (open circuit)).

This procedure is followed for each source in turn, then the resultant responses are added to determine the true operation of the circuit. The resultant circuit operation is the superposition of the various voltage and current sources.

The superposition theorem is very important in circuit analysis. It is used in converting any circuit into its Norton equivalent or Thevenin equivalent.

The theorem is applicable to linear networks (time varying or time invariant) consisting of independent sources, linear dependent sources, linear passive elements (resistors, inductors, capacitors) and linear transformers.

Superposition works for voltage and current but not power. In other words, the sum of the powers of each source with the other sources turned off is not the real consumed power. To calculate power we first use superposition to find both current and voltage of each linear element and then calculate the sum of the multiplied voltages and currents.

Gas Pressure Analogy

The electric circuit superposition theorem is analogous to Dalton's law of partial pressure which can be stated as the total pressure exerted by an ideal gas mixture in a given volume is the algebraic sum of all the pressures exerted by each gas if it were alone in that volume.

RECIPROCITY THEOREM

In its simplest form, the reciprocity theorem states that if an emf E in one branch of a reciprocal network produces a current I in another, then if the emf E is moved from the first to the second branch, it will cause the same current in the first branch, where the emf has been replaced by a short circuit.

Any network composed of linear, bilateral elements (such as R, L and C) is reciprocal.

The circuit in the figure is a concrete example of reciprocity. The reader should solve the circuit, and determine the values of the current I in the two cases, which will be equal (0.35294 A). If E is reversed, then the direction of I is reversed, so the direction does not matter so long as both E and I are reversed at the same time.

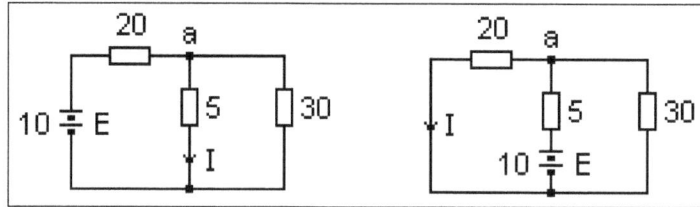

Illustrating reciprocity.

A non-bilateral element, such as a rectifying diode, destroys reciprocity, as the circuit at the left shows. Here it is obvious that I = 0 on the right, while I ≠ 0 on the left. Even if the polarity of E is reversed so that the diode is forward-biased on the right, the currents are not reciprocal. On the left, I will be 0.21294 A, while on the right it will be 0.32824 A, as the reader can show by solving the circuits.

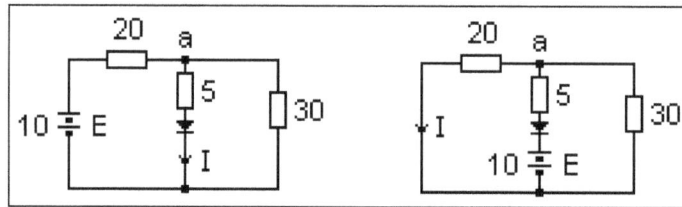

Non-bilateral element.

A nonlinear element also destroys reciprocity. The circuit at the right includes a resistor whose voltage drop is proportional to the square of the current, $V = 10i^2$. In the circuit at the left, I = 0.3798 A, while on the right I = 0.3397 A. The inclusion of controlled sources or active elements may also destroy reciprocity.

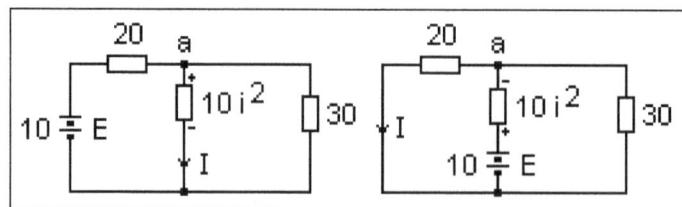

Nonlinear element.

Two-Port Networks

Consideration of reciprocity leads naturally to two-port networks. These are networks with four terminals considered in two pairs as *ports* at which connections are made. The emf E in the reciprocity theorem is considered to be connected to one port, say port 1, while the current is at port

2, assumed to be short-circuited. The ports result from breaking into two of the branches of the network. One terminal of each port is denoted by (+) to specify the polarity of the voltage applied at the port, and currents are positive when they enter the (+) terminal.

The fundamental variables are V_1, I_1, V_2 and I_2. Any two of these variables are functions of the remaining two. For certain networks, some of the four choices are not admissible. In most cases, the variables appearing in the models are variations from DC bias conditions, not the DC variables themselves.

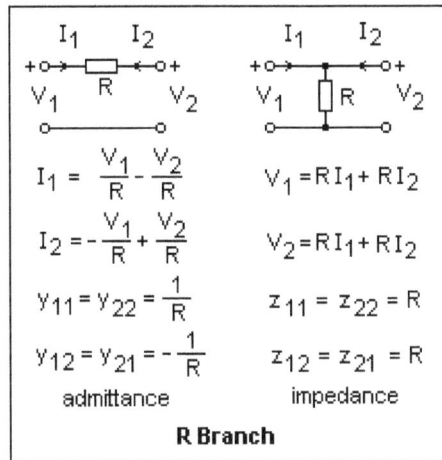

R Branch

A single resistor forms two two-ports, depending on whether it is in series or shunt. For the series resistor, it is normal to take the dependent variables as I_1 and I_2, and the independent variables V_1 and V_2. The coefficients are called the *admittance parameters*, since admittance is the ratio of current to voltage. If the resistor is connected in shunt, the natural independent variables are I_1 and $_2$, while V_1 and V_2 are the dependent variables. The coefficients in this case are the *impedance parameters*, since impedance is the ratio of voltage to current. In both cases, we see that the off-diagonal or *transfer* coefficients are equal.

Consider the admittance model, where the coefficients are not necessarily equal to the simple ones for a single resistor. If an emf E is applied at the input, $E = V_1$, and the output is shorted, $V_2 = 0$, we see that $I_2 = y_{21}E$. If we apply the same voltage E to the output, and short the input, we have $I_1 = y_{12}E$. Therefore, the currents are equal, and the network is *reciprocal*. Reciprocity is the result of the equality of the transfer admittances. A similar result can be obtained for the impedance model, where reciprocity holds if $z_{12} = z_{21}$, as the reader can easily show. For our simple example of a single resistor, these conditions hold.

We note that there is no impedance model for the series resistor, and no admittance model for the shunt resistor. In general, it is possible to transform from impedance to admittance and vice versa, but the determinant of the coefficients is in the denominator of the transformation formula, and for a single resistor this determinant is zero. If Δ is the determinant of the admittance coefficients, then $z_{11} = y_{22}/\Delta$, $z_{22} = y_{11}/\Delta$, $z_{12} = y_{21}/\Delta$ and $z_{21} = y_{12}/\Delta$. Similar formulas hold for the inverse transformation. A full list of transformations is given in the first Reference. The conclusion is that if $y_{12} = y_{21}$, then $z_{21} = z_{12}$ as well. This, of course, is a general result. The particular way of modeling a network does not affect the property of reciprocity.

A popular model, especially for active elements, is the *hybrid* model in which the independent variables are V_2 and I_1. The equations for the series resistor are $V_1 = RI_1 + V_2$ and $I_2 = -I_1$. The hybrid parameters for this case are $h_{11} = R$, $h_{22} = 0$, $h_{12} = -h_{21} = 1$. Note that the off-diagonal elements are of opposite sign in this case, but of the same magnitude. This is the mark of reciprocity in the hybrid model. As an exercise, the reader may show that for the shunt resistor $h_{11} = 0$, $h_{22} = 1/R$, and the transfer coefficients are the same as those for the series resistor.

There is a fourth model in which the output variables I_1 and V_1 are expressed in terms of the input variables V_2 and I_2: $V_1 = AV_2 - BI_2$, and $I_1 = CV_2 - DI_2$. Note the change in sign of I_2, which is conventional so it will be in the same direction as I_1 for an identity transformation $A = D = 1$, $B = C = 0$. If networks are cascaded, these matrices simply multiply. In this model, reciprocity is expressed by $AD - CB = 1$. This can be proved by expressing A, B, C and D in terms of the impedance paramters. $A = z_{11}/z_{21}$, $B = z_{11}z_{22}/z_{21} - z_{12}$, $C = 1/z_{21}$ and $D = z_{22}/z_{21}$. The result is $AD - CB = z_{12}/z_{21} = 1$ if the network is reciprocal. It is also clear that any cascade of reciprocal networks is also reciprocal, since the determinant of the product of matrices is the product of the determinants.

Any model can be expressed in terms of controlled sources and impedances. This is shown at the right for they hybrid model. The h_{11} is represented by an impedance, h_{22} by an admittance. h_{12} and h_{21} are the reverse and forward transfer ratios, which are pure numbers. A current source appears in the output, while a voltage source appears in the input. This is the model often used to represent a transistor, in which h_{21} is usually denoted h_{fe} or β, the base to collector current gain. The input is between base and emitter, the output between collector and emitter. Since h_{12} is much smaller than h_{21}, the transistor is not reciprocal. A good approximate model results if $h_{11} = h_{21}(25\Omega/I_c)$, $h_{12} = h_{22} = 0$. I_c is the DC collector current. A typical value for h_{21} is 100. For accurate work, the hybrid coefficients can be expressed as functions of the bias conditions.

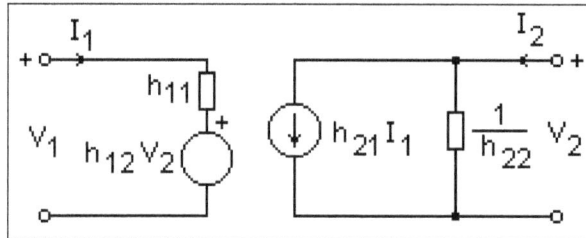

Hybrid model.

A vacuum tube can be represented by an admittance model, in which the plate current I_p is a joint function of the grid voltage V_g and the plate voltage V_p. The input port is connected to grid and cathode, the output port to plate and cathode. As long as the grid is biased negatively with respect to the cathode, $I_g = 0$, and two of the admittance parameters are zero. The remaining equation can be written $I_p = g_m V_g + V_p/r_p$, where $y_{12} = g_m$ is the *transconductance* and $1/y_{22} = r_p$ is the *plate resistance*. When I_p is held constant, $-g_m r_p V_g = V_p$, which defines the *amplification factor* $\mu = g_m r_p$. The same model can be used for an FET. We note that none of these models are reciprocal, which is worth remembering. These models are extremely useful in circuit design.

Proof of the Reciprocity Theorem

We wish to show that in a network of linear, bilinear elements, that is, in one constructed of of ordinary impedances, that if when a voltage V is inserted in one loop the current I in another loop due to the insertion of this voltage is the same as the current at the first position due to the insertion of a voltage V in the second loop, or that the network is *reciprocal*. We shall do this by explicitly calculating the currents in the two cases, and observing that they are equal.

Consider the network as made up of N independent loops, of which loop 1 contains the input port, and loop 2 the output port. Let there be no emf's in the network. If there are, they can be taken care of by superposition, and we can set them all to zero for our purposes. The loop currents are I_k, k = 1 to N. If emf V is at port 1 and port 2 is shorted, then Kirchhoff's laws give $\Sigma z_{1k}I_k = V$, and $\Sigma z_{jk}I_k = 0$, j = 2 to N. The current I_2 can then be expressed in terms of determinants as $I_2 = -V \det A / \Delta$, where the determinant in the numerator has been expanded by minors of the second column, which is all zeros except for V in the first position. A is the matrix of the z's with the first row and second column taken away. Δ is the determinant of the matrix of the z_{jk}. Write out this solution in detail if it is not clear from this description.

Now connect V in port 2, and short port 1. The solution by determinants for I_1 is -V det B/Δ, where B is the matrix of the z's with the second row and first column taken away. If we compare matrices A and B, we see that they are transposes of each other, provided that $z_{jk} = z_{kj}$. However, the off-diagonal z's are just the mutual impedances of the current loops (impedances common to a pair of loops), and do not depend on the order of the subscripts. (They are not the coefficients of an impedance model here, simply actual impedances.) Since the determinant of the transpose of a matrix is equal to the determinant of the matrix, the two solutions are the same, and $I_1 = I_2$. This proves the reciprocity theorem for networks made from linear, bilateral elements, Q.E.D.

We have proved the reciprocity theorem for a relatively limited class of networks, but it is possible to extend the theorem more widely.

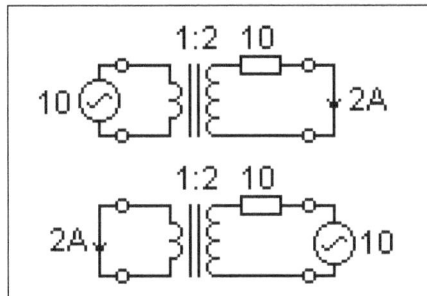

Transformers are reciprocal.

The figure at the above shows that reciprocity holds for ideal transformers at least.

Since real transformers may be modeled by networks containing only impedances and ideal transformers, real transformers must also be reciprocal, and this extends the utility of the theorem further. Reciprocity crops up in many unexpected places, such as in electromagnetic fields and microwaves.

EXTRA ELEMENT THEOREM

The Extra Element Theorem (EET) is an analytic technique developed by R. D. Middlebrook for simplifying the process of deriving driving point and transfer functions for linear electronic circuits. Much like Thévenin's theorem, the extra element theorem breaks down one complicated problem into several simpler ones.

Driving point and transfer functions can generally be found using KVL and KCL methods, however several complicated equations may result that offer little insight into the circuit's behavior. Using the extra element theorem, a circuit element (such as a resistor) can be removed from a circuit and the desired driving point or transfer function found. By removing the element that most complicates the circuit (such as an element that creates feedback), the desired function can be easier to obtain. Next two correctional factors must be found and combined with the previously derived function to find the exact expression.

The general form of the extra element theorem is called the N-extra element theorem and allows multiple circuit elements to be removed at once.

The (single) extra element theorem expresses any transfer function as a product of the transfer function with that element removed and a correction factor. The correction factor term consists of the impedance of the extra element and two driving point impedances seen by the extra element: The double null injection driving point impedance and the single injection driving point impedance. Because an extra element can be removed in general by either short-circuiting or open-circuiting the element, there are two equivalent forms of the EET:

$$H(s) = H_\infty(s)\frac{1+\dfrac{Z_n(s)}{Z(s)}}{1+\dfrac{Z_d(s)}{Z(s)}}$$

or,

$$H(s) = H_0(s)\frac{1+\dfrac{Z(s)}{Z_n(s)}}{1+\dfrac{Z(s)}{Z_d(s)}}$$

Where the Laplace-domain transfer functions and impedances in the above expressions are defined as follows: $H(s)$ is the transfer function with the extra element present. $H_\infty(s)$ is the transfer function with the extra element open-circuited. $H_0(s)$ is the transfer function with the extra element short-circuited. $Z(s)$ is the impedance of the extra element. $Z_d(s)$ is the single-injection driving point impedance "seen" by the extra element. $Z_n(s)$ is the double-null-injection driving point impedance "seen" by the extra element.

The extra element theorem incidentally proves that any electric circuit transfer function can be expressed as no more than a bilinear function of any particular circuit element.

Driving Point Impedances

Single Injection Driving Point Impedance

$Z_d(s)$ is found by making the input to the system's transfer function zero (short circuit a voltage source or open circuit a current source) and determining the impedance across the terminals to which the extra element will be connected with the extra element absent. This impedance is same as the Thévenin's equivalent impedance.

Double Null Injection Driving Point Impedance

$Z_n(s)$ is found by replacing the extra element with a second test signal source (either current source or voltage source as appropriate). Then, $Z_n(s)$ is defined as the ratio of voltage across the terminals of this second test source to the current leaving its positive terminal when the output of the system's transfer function is nulled for any value of the primary input to the system's transfer function.

In practice, $Z_n(s)$ can be found from working backwards from the facts that the output of the transfer function is made zero and that the primary input to the transfer function is unknown. Then using conventional circuit analysis techniques to express both the voltage across the extra element test source's terminals, $v_n(s)$, and the current leaving the extra element test source's positive terminals, $i_n(s)$, and calculating $Z_n(s) = v_n(s)/i_n(s)$. Although computation of $Z_n(s)$ is an unfamiliar process for many engineers, its expressions are often much simpler than those for $Z_d(s)$ because the nulling of the transfer function's output often leads to other voltages/currents in the circuit being zero, which may allow exclusion of certain components from analysis.

Special Case With Transfer Function as a Self-Impedance

As a special case, the EET can be used to find the input impedance of a network with the addition of an element designated as "extra". In this case, Z_d is same as the impedance of the input test current source signal made zero or equivalently with the input open circuited. Likewise, since the transfer function output signal can be considered to be the voltage at the input terminals, Z_n is found when the input voltage is zero i.e. the input terminals are short-circuited. Thus, for this particular application the EET can be written as:

$$Z_{in} = Z_{in}^{\infty} \cdot \frac{1 + \dfrac{Z_e^0}{Z}}{1 + \dfrac{Z_e^{\infty}}{Z}}$$

where,

Z is the impedance chosen as the extra element.

Z_{in}^{∞} is the input impedance with Z removed (or made infinite).

Z_e^0 is the impedance seen by the extra element Z with the input shorted (or made zero).

Z_e^{∞} is the impedance seen by the extra element Z with the input open (or made infinite).

Computing these three terms may seem like extra effort, but they are often easier to compute than the overall input impedance.

Example:

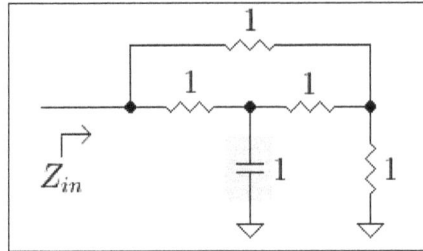

Simple RC circuit to demonstrate the EET. The capacitor (gray shading) is denoted the extra element.

Consider the problem of finding Z_{in} for the circuit in Figure using the EET (note all component values are unity for simplicity). If the capacitor (gray shading) is denoted the extra element then,

$$Z = \frac{1}{s}$$

Removing this capacitor from the circuit we find,

$$Z_{in}^{\infty} = 2 \| 1 + 1 = \frac{5}{3}$$

Calculating the impedance seen by the capacitor with the input shorted we find,

$$Z_{e}^{0} = 1 \| (1 + 1 \| 1) = \frac{3}{5}$$

Calculating the impedance seen by the capacitor with the input open we find,

$$Z_{e}^{\infty} = 2 \| 1 + 1 = \frac{5}{3}$$

Therefore, using the EET, we find,

$$Z_{in} = \frac{5}{3} \cdot \frac{1 + \frac{3}{5}s}{1 + \frac{5}{3}s} = \frac{5 + 3s}{3 + 5s}$$

Note that this problem was solved by calculating three simple driving point impedances by inspection.

MAXIMUM POWER TRANSFER THEOREM

In electrical engineering, the maximum power transfer theorem states that, to obtain *maximum* external power from a source with a finite internal resistance, the resistance of the load must equal the resistance of the source as viewed from its output terminals. Moritz von Jacobi published the maximum power (transfer) theorem around 1840; it is also referred to as "Jacobi's law".

The theorem results in maximum *power* transfer across the circuit, and not maximum *efficiency*. If the resistance of the load is made larger than the resistance of the source, then efficiency is higher, since a higher percentage of the source power is transferred to the load, but the *magnitude* of the load power is lower since the total circuit resistance goes up.

If the load resistance is smaller than the source resistance, then most of the power ends up being dissipated in the source, and although the total power dissipated is higher, due to a lower total resistance, it turns out that the amount dissipated in the load is reduced.

The theorem states how to choose (so as to maximize power transfer) the load resistance, once the source resistance is given. It is a common misconception to apply the theorem in the opposite scenario. It does *not* say how to choose the source resistance for a given load resistance. In fact, the source resistance that maximizes power transfer is always zero, regardless of the value of the load resistance.

The theorem can be extended to alternating current circuits that include reactance, and states that maximum power transfer occurs when the load impedance is equal to the complex conjugate of the source impedance.

Maximizing Power Transfer versus Power Efficiency

The theorem was originally misunderstood (notably by Joule) to imply that a system consisting of an electric motor driven by a battery could not be more than 50% efficient since, when the impedances were matched, the power lost as heat in the battery would always be equal to the power delivered to the motor.

In 1880 this assumption was shown to be false by either Edison or his colleague Francis Robbins Upton, who realized that maximum efficiency was not the same as maximum power transfer.

To achieve maximum efficiency, the resistance of the source (whether a battery or a dynamo) could be (or should be) made as close to zero as possible. Using this new understanding, they obtained an efficiency of about 90%, and proved that the electric motor was a practical alternative to the heat engine.

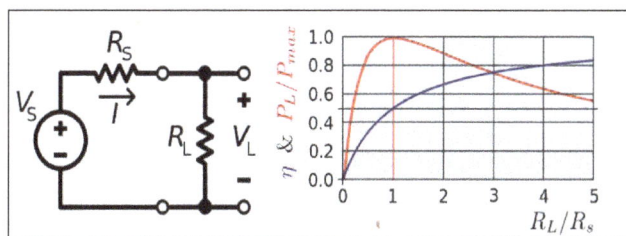

The condition of maximum power transfer does not result in maximum efficiency.

If we define the efficiency η as the ratio of power dissipated by the load, R_L, to power developed by the source, V_S, then it is straightforward to calculate from the above circuit diagram that,

$$\eta = \frac{R_L}{R_L + R_S} = \frac{1}{1 + R_S / R_L}.$$

Consider three particular cases:

- If $R_L = R_S$, then $\eta = 0.5$,
- If $R_L \rightarrow \infty$ or $R_S = 0$, then $\eta = 1$,
- If $R_L = 0$, then $\eta = 0$.

The efficiency is only 50% when maximum power transfer is achieved, but approaches 100% as the load resistance approaches infinity, though the total power level tends towards zero.

Efficiency also approaches 100% if the source resistance approaches zero, and 0% if the load resistance approaches zero. In the latter case, all the power is consumed inside the source (unless the source also has no resistance), so the power dissipated in a short circuit is zero.

Impedance Matching

A related concept is reflectionless impedance matching.

In radio frequency transmission lines, and other electronics, there is often a requirement to match the source impedance (at the transmitter) to the load impedance (such as an antenna) to avoid reflections in the transmission line that could overload or damage the transmitter.

Calculus-based Proof for Purely Resistive Circuits

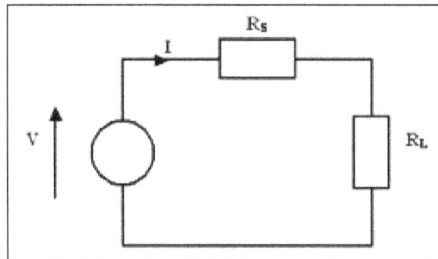

In the diagram opposite, power is being transferred from the source, with voltage V and fixed source resistance R_S, to a load with resistance R_L, resulting in a current I. By Ohm's law, I is simply the source voltage divided by the total circuit resistance:

$$I = \frac{V}{R_S + R_L}.$$

The power P_L dissipated in the load is the square of the current multiplied by the resistance:

$$P_L = I^2 R_L = \left(\frac{V}{R_S + R_L}\right)^2 R_L = \frac{V^2}{R_S^2 / R_L + 2R_S + R_L}.$$

The value of R_L for which this expression is a maximum could be calculated by differentiating it, but it is easier to calculate the value of R_L for which the denominator,

$$R_S^2 / R_L + 2R_S + R_L$$

is a minimum. The result will be the same in either case. Differentiating the denominator with respect to R_L:

$$\frac{d}{dR_L}\left(R_S^2 / R_L + 2R_S + R_L\right) = -R_S^2 / R_L^2 + 1.$$

For a maximum or minimum, the first derivative is zero, so

$$R_S^2 / R_L^2 = 1$$

or

$$R_L = \pm R_S.$$

In practical resistive circuits, R_S and R_L are both positive, so the positive sign in the above is the correct solution.

To find out whether this solution is a minimum or a maximum, the denominator expression is differentiated again:

$$\frac{d^2}{dR_L^2}\left(R_S^2 / R_L + 2R_S + R_L\right) = 2R_S^2 / R_L^3.$$

This is always positive for positive values of R_S and R_L , showing that the denominator is a minimum, and the power is therefore a maximum, when

$$R_S = R_L.$$

The above proof assumes fixed source resistance R_S. When the source resistance can be varied, power transferred to the load can be increased by reducing R_S. For example, a 100 Volt source with an R_S of 10Ω will deliver 250 watts of power to a 10Ω load; reducing R_S to 0Ω increases the power delivered to 1000 watts.

Note that this shows that maximum power transfer can also be interpreted as the load voltage being equal to one-half of the Thevenin voltage equivalent of the source.

In Reactive Circuits

The power transfer theorem also applies when the source and/or load are not purely resistive.

A refinement of the maximum power theorem says that any reactive components of source and load should be of equal magnitude but opposite sign.

- This means that the source and load impedances should be *complex conjugates* of each other.

- In the case of purely resistive circuits, the two concepts are identical.

Physically realizable sources and loads are not usually purely resistive, having some inductive or capacitive components, and so practical applications of this theorem, under the name of complex conjugate impedance matching, do, in fact, exist.

If the source is totally inductive (capacitive), then a totally capacitive (inductive) load, in the absence of resistive losses, would receive 100% of the energy from the source but send it back after a quarter cycle.

The resultant circuit is nothing other than a resonant LC circuit in which the energy continues to oscillate to and from. This oscillation is called reactive power.

Power factor correction (where an inductive reactance is used to "balance out" a capacitive one), is essentially the same idea as complex conjugate impedance matching although it is done for entirely different reasons.

For a fixed reactive *source*, the maximum power theorem maximizes the real power (P) delivered to the load by complex conjugate matching the load to the source.

For a fixed reactive *load*, power factor correction minimizes the apparent power (S) (and unnecessary current) conducted by the transmission lines, while maintaining the same amount of real power transfer.

This is done by adding a reactance to the load to balance out the load's own reactance, changing the reactive load impedance into a resistive load impedance.

Proof

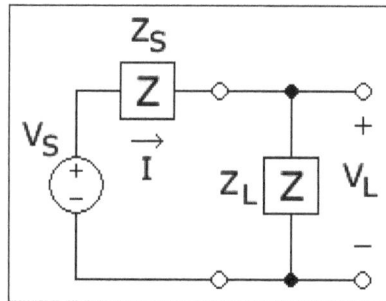

In this diagram, AC power is being transferred from the source, with phasor magnitude of voltage $|V_S|$ (positive peak voltage) and fixed source impedance Z_S (S for source), to a load with impedance Z_L (L for load), resulting in a (positive) magnitude $|I|$ of the current phasor I. This magnitude $|I|$ results from dividing the magnitude of the source voltage by the magnitude of the total circuit impedance:

$$|I| = \frac{|V_S|}{|Z_S + Z_L|}.$$

The average power P_L dissipated in the load is the square of the current multiplied by the resistive portion (the real part) R_L of the load impedance Z_L:

$$P_L = I_{rms}^2 R_L = \frac{1}{2}|I|^2 R_L = \frac{1}{2}\left(\frac{|V_S|}{|Z_S + Z_L|}\right)^2 R_L = \frac{1}{2}\frac{|V_S|^2 R_L}{(R_S + R_L)^2 + (X_S + X_L)^2},$$

where, and R_L denote the resistances, that is the real parts, and X_S and X_L denote the reactances, that is the imaginary parts, of respectively the source and load impedances Z_S and Z_L.

To determine, for a given source voltage V_S and impedance Z_S, the value of the load impedance Z_L, for which this expression for the power yields a maximum, one first finds, for each fixed positive value of R_L, the value of the reactive term X_L for which the denominator

$$(R_S + R_L)^2 + (X_S + X_L)^2$$

is a minimum. Since reactances can be negative, this is achieved by adapting the load reactance to,

$$X_L = -X_S.$$

This reduces the above equation to:

$$P_L = \frac{1}{2} \frac{|V_S|^2 R_L}{(R_S + R_L)^2}$$

and it remains to find the value of R_L which maximizes this expression. This problem has the same form as in the purely resistive case, and the maximizing condition therefore is $R_L = R_S$.

The two maximizing conditions:

- $R_L = R_S.$

- $X_L = -X_S.$

describe the complex conjugate of the source impedance, denoted by and thus can be concisely combined to:

$$Z_L = Z_S^*.$$

COMPENSATION THEOREM

Compensation Theorem states that in a linear time invariant network when the resistance (R) of an uncoupled branch, carrying a current (I), is changed by (ΔR). The currents in all the branches would change and can be obtained by assuming that an ideal voltage source of (V_C) has been connected such that V_C = I (ΔR) in series with (R + ΔR) when all other sources in the network are replaced by their internal resistances.

In Compensation Theorem, the source voltage (V_C) opposes the original current. In simple words compensation theorem can be stated as – the resistance of any network can be replaced by a voltage source, having the same voltage as the voltage drop across the resistance which is replaced.

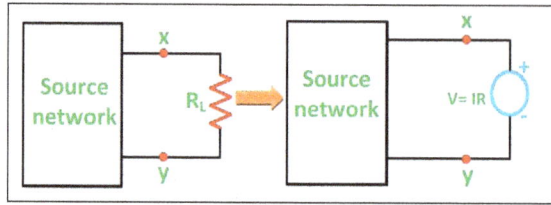

Let us assume a load R_L be connected to a DC source network whose Thevenin's equivalent gives V_0 as the Thevenin's voltage and R_{TH} as the Thevenin's resistance as shown in the figure below:

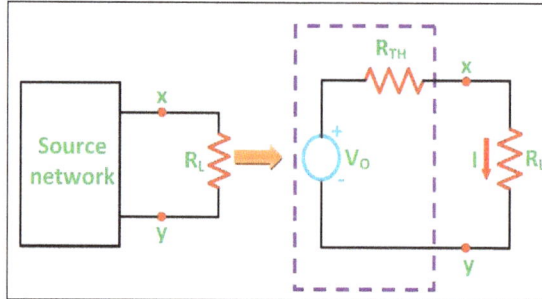

Here,

$$I=\frac{V_0}{R_{TH}+R_L}$$

Let the load resistance RL be changed to (RL + ΔRL). Since the rest of the circuit remains unchanged, the Thevenin's equivalent network remains the same as shown in the circuit diagram below:

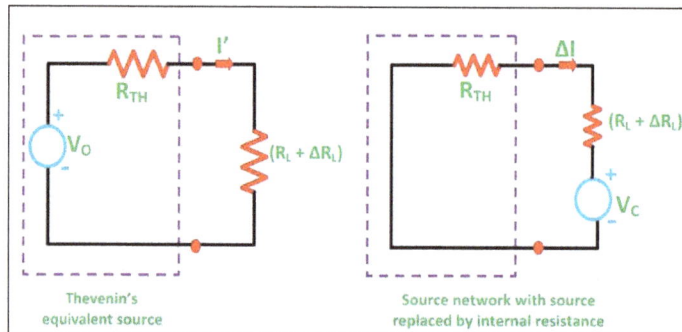

Here,

$$I'=\frac{V_0}{R_{TH}+(R_L+\Delta R_L)}$$

The change of current being termed as ΔI.

Therefore,

$$\Delta I = I' - I$$

Putting the value of I' and I from the equation $I = \dfrac{V_0}{R_{TH} + R_L}$ and $I' = \dfrac{V_0}{R_{TH} + (R_L + \Delta R_L)}$ in the equation $\Delta I = I' - I$ we will get the following equation:

$$\Delta I = \frac{V_0}{R_{TH} + (R_L + \Delta R_L)} - \frac{V_0}{R_{TH} + R_L}$$

$$\Delta I = \frac{V_0 \{R_{TH} + R_L - (R_{TH} + R_L + \Delta R_L)\}}{(R_{TH} + R_L + \Delta R_L)(R_{TH} + R_L)}$$

$$\Delta I = -\left[\frac{V_0}{R_{TH} + R_L}\right] \frac{\Delta R_L}{R_{TH} + R_L + \Delta R_L}$$

Now, putting the value of I from the equation $I = \dfrac{V_0}{R_{TH} + R_L}$ in the above equation, we will get the following equation

$$\Delta I = -\frac{I \Delta R_L}{R_{TH} + R_L + \Delta R_L}$$

As we know, VC = I Δ RL and is known as compensating voltage.

Therefore, the equation $\Delta I + -\dfrac{I \Delta R_L}{R_{TH} + R_L + \Delta R_L}$ becomes,

$$\Delta I = \frac{-V_C}{R_{TH} + R_L + \Delta R_L}$$

Hence, Compensation Theorem tells that with the change of branch resistance, branch currents changes and the change is equivalent to an ideal compensating voltage source in series with the branch opposing the original current, all other sources in the network being replaced by their internal resistances.

Y-Δ TRANSFORM

The Y-Δ transform, also written wye-delta and also known by many other names, is a mathematical technique to simplify the analysis of an electrical network. The name derives from the shapes of the circuit diagrams, which look respectively like the letter Y and the Greek capital letter Δ. This circuit transformation theory was published by Arthur Edwin Kennelly in 1899. It is widely used in analysis of three-phase electric power circuits.

The Y-Δ transform can be considered a special case of the star-mesh transform for three resistors. In mathematics, the Y-Δ transform plays an important role in theory of circular planar graphs.

Names

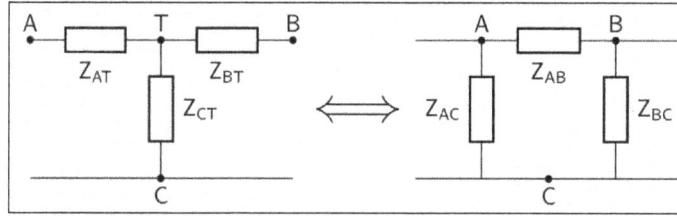

Illustration of the transform in its T-Π representation.

The Y-Δ transform is known by a variety of other names, mostly based upon the two shapes involved, listed in either order. The Y, spelled out as wye, can also be called T or star; the Δ, spelled out as delta, can also be called triangle, Π (spelled out as pi), or mesh. Thus, common names for the transformation include wye-delta or delta-wye, star-delta, star-mesh, or T-Π.

Basic Y-Δ Transformation

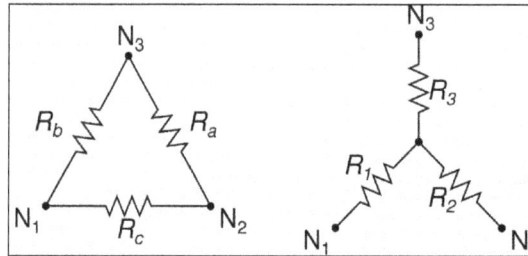

Δ and Y circuits with the labels which are used in this article.

The transformation is used to establish equivalence for networks with three terminals. Where three elements terminate at a common node and none are sources, the node is eliminated by transforming the impedances. For equivalence, the impedance between any pair of terminals must be the same for both networks. The equations given here are valid for complex as well as real impedances.

Equations for the Transformation from Δ to Y

The general idea is to compute the impedance R_Y at a terminal node of the Y circuit with impedances R', R'' to adjacent nodes in the Δ circuit by,

$$R_Y = \frac{R'R''}{\sum R_\Delta}$$

where R_Δ are all impedances in the Δ circuit. This yields the specific formulae,

$$R_1 = \frac{R_b R_c}{R_a + R_b + R_c}$$

$$R_2 = \frac{R_a R_c}{R_a + R_b + R_c}$$

$$R_3 = \frac{R_a R_b}{R_a + R_b + R_c}$$

Equations for the Transformation from Y to Δ

The general idea is to compute an impedance R_Δ in the Δ circuit by,

$$R_\Delta = \frac{R_P}{R_{\text{opposite}}}$$

where $R_P = R_1 R_2 + R_2 R_3 + R_3 R_1$ is the sum of the products of all pairs of impedances in the Y circuit and R_{opposite} is the impedance of the node in the Y circuit which is opposite the edge with R_Δ. The formulae for the individual edges are thus,

$$R_a = \frac{R_1 R_2 + R_2 R_3 + R_3 R_1}{R_1}$$

$$R_b = \frac{R_1 R_2 + R_2 R_3 + R_3 R_1}{R_2}$$

$$R_c = \frac{R_1 R_2 + R_2 R_3 + R_3 R_1}{R_3}$$

Or, if using admittance instead of resistance:

$$Y_a = \frac{Y_3 Y_2}{\sum Y_Y}$$

$$Y_b = \frac{Y_3 Y_1}{\sum Y_Y}$$

$$Y_c = \frac{Y_1 Y_2}{\sum Y_Y}$$

Note that the general formula in Y to Δ using admittance is similar to Δ to Y using resistance.

A Proof of the Existence and Uniqueness of the Transformation

The feasibility of the transformation can be shown as a consequence of the superposition theorem for electric circuits. A short proof, rather than one derived as a corollary of the more general star-mesh transform, can be given as follows. The equivalence lies in the statement that for any external voltages (V_1, V_2 and V_3) applying at the three nodes (N_1, N_2 and N_3), the corresponding currents (I_1, I_2 and I_3) are exactly the same for both the Y and Δ circuit, and vice versa. In this proof, we start with given external currents at the nodes. According to the superposition theorem, the voltages can be obtained by studying the superposition of the resulting voltages at the nodes of the following three problems applied at the three nodes with current:

- $\frac{1}{3}(I_1 - I_2), -\frac{1}{3}(I_1 - I_2), 0$

- $0, \frac{1}{3}(I_2 - I_3), -\frac{1}{3}(I_2 - I_3)$ and

- $-\dfrac{1}{3}(I_3 - I_1), 0, \dfrac{1}{3}(I_3 - I_1)$

The equivalence can be readily shown by using Kirchhoff's circuit laws that $I_1 + I_2 + I_3 = 0$. Now each problem is relatively simple, since it involves only one single ideal current source. To obtain exactly the same outcome voltages at the nodes for each problem, the equivalent resistances in the two circuits must be the same, this can be easily found by using the basic rules of series and parallel circuits:

$$R_3 + R_1 = \frac{(R_c + R_a)R_b}{R_a + R_b + R_c}, \quad \frac{R_3}{R_1} = \frac{R_a}{R_c}.$$

Though usually six equations are more than enough to express three variables (R_1, R_2, R_3) in term of the other three variables(R_a, R_b, R_c), here it is straightforward to show that these equations indeed lead to the above designed expressions.

In fact, the superposition theorem establishes the relation between the values of the resistances, the uniqueness theorem guarantees the uniqueness of such solution.

Simplification of Networks

Resistive networks between two terminals can theoretically be simplified to a single equivalent resistor (more generally, the same is true of impedance). Series and parallel transforms are basic tools for doing so, but for complex networks such as the bridge illustrated here, they do not suffice.

The Y-Δ transform can be used to eliminate one node at a time and produce a network that can be further simplified, as shown.

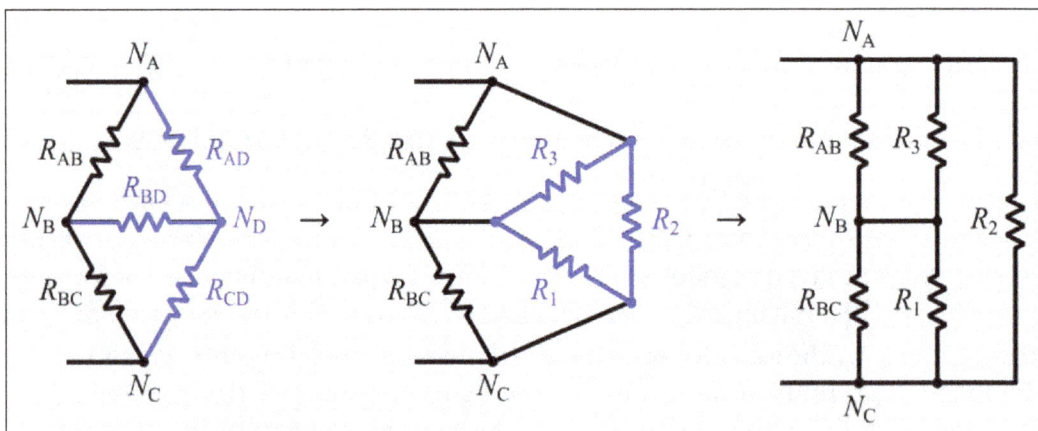

Transformation of a bridge resistor network, using the Y-Δ transform to eliminate node *D*, yields an equivalent network that may readily be simplified further.

The reverse transformation, Δ-Y, which adds a node, is often handy to pave the way for further simplification as well.

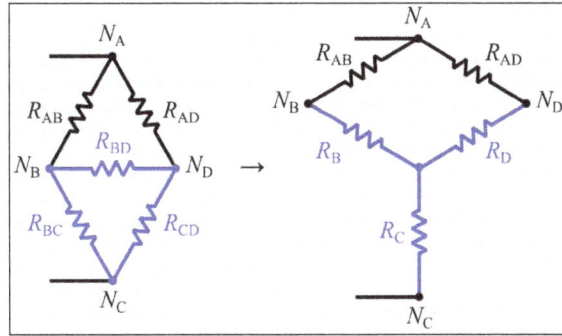

Transformation of a bridge resistor network, using the Δ-Y transform, also yields an equivalent network that may readily be simplified further.

Every two-terminal network represented by a planar graph can be reduced to a single equivalent resistor by a sequence of series, parallel, Y-Δ, and Δ-Y transformations. However, there are non-planar networks that cannot be simplified using these transformations, such as a regular square grid wrapped around a torus, or any member of the Petersen family.

Graph Theory

In graph theory, the Y-Δ transform means replacing a Y subgraph of a graph with the equivalent Δ subgraph. The transform preserves the number of edges in a graph, but not the number of vertices or the number of cycles. Two graphs are said to be Y-Δ equivalent if one can be obtained from the other by a series of Y-Δ transforms in either direction. For example, the Petersen family is a Y-Δ equivalence class.

Demonstration

Δ-load to Y-load Transformation Equations

To relate $\{R_a, R_b, R_c\}$ from Δ to $\{R_1, R_2, R_3\}$ from Y, the impedance between two corresponding nodes is compared. The impedance in either configuration is determined as if one of the nodes is disconnected from the circuit.

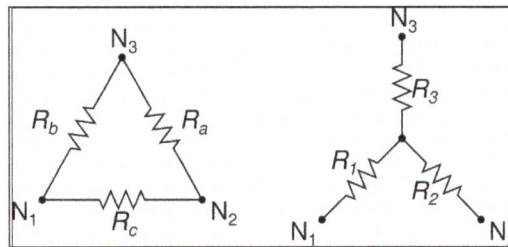

Δ and Y circuits with the labels.

The impedance between N_1 and N_2 with N_3 disconnected in Δ:

$$R_\Delta(N_1, N_2) = R_c \parallel (R_a + R_b)$$

$$= \cfrac{1}{\cfrac{1}{R_c} + \cfrac{1}{R_a + R_b}}$$

$$= \frac{R_c\left(R_a + R_b\right)}{R_a + R_b + R_c}$$

To simplify, let R_T be the sum of $\{R_a, R_b, R_c\}$.

$$R_T = R_a + R_b + R_c$$

Thus,

$$R_\Delta\left(N_1, N_2\right) = \frac{R_c\left(R_a + R_b\right)}{R_T}$$

The corresponding impedance between N_1 and N_2 in Y is simple:

$$R_Y\left(N_1, N_2\right) = R_1 + R_2$$

hence:

$$R_1 + R_2 = \frac{R_c\left(R_a + R_b\right)}{R_T}$$

Repeating for $R(N_2, N_3)$:

$$R_2 + R_3 = \frac{R_a\left(R_b + R_c\right)}{R_T}$$

and for $R\left(N_1, N_3\right)$:

$$R_1 + R_3 = \frac{R_b\left(R_a + R_c\right)}{R_T}.$$

From here, the values of $\{R_1, R_2, R_3\}$ can be determined by linear combination (addition and/or subtraction).

For example, adding $R_1 + R_2 = \dfrac{R_c\left(R_a + R_b\right)}{R_T}$ and $R_1 + R_3 = \dfrac{R_b\left(R_a + R_c\right)}{R_T}.$, then subtracting

$R_2 + R_3 = \dfrac{R_a\left(R_b + R_c\right)}{R_T}$ yields

$$R_1 + R_2 + R_1 + R_3 - R_2 - R_3 = \frac{R_c\left(R_a + R_b\right)}{R_T} + \frac{R_b\left(R_a + R_c\right)}{R_T} - \frac{R_a\left(R_b + R_c\right)}{R_T}$$

$$\Rightarrow 2R_1 = \frac{2R_b R_c}{R_T}$$

$$\Rightarrow R_1 = \frac{R_b R_c}{R_T}.$$

For completeness:

$$R_1 = \frac{R_b R_c}{R_T}$$

$$R_2 = \frac{R_a R_c}{R_T}$$

$$R_3 = \frac{R_a R_b}{R_T}$$

Y-load to Δ-load Transformation Equations

Let,

$$R_T = R_a + R_b + R_c.$$

We can write the Δ to Y equations as,

$$R_1 = \frac{R_b R_c}{R_T}$$

$$R_2 = \frac{R_a R_c}{R_T}$$

$$R_3 = \frac{R_a R_b}{R_T}.$$

Multiplying the pairs of equations yields,

$$R_1 R_2 = \frac{R_a R_b R_c^2}{R_T^2}$$

$$R_1 R_3 = \frac{R_a R_b^2 R_c}{R_T^2}$$

$$R_2 R_3 = \frac{R_a^2 R_b R_c}{R_T^2}$$

and the sum of these equations is,

$$R_1 R_2 + R_1 R_3 + R_2 R_3 = \frac{R_a R_b R_c^2 + R_a R_b^2 R_c + R_a^2 R_b R_c}{R_T^2}$$

Factor $R_a R_b R_c$ from the right side, leaving R_T in the numerator, canceling with an R_T in the denominator.

$$R_1R_2 + R_1R_3 + R_2R_3 = \frac{(R_aR_bR_c)(R_a + R_b + R_c)}{R_T^2}$$

$$= \frac{R_aR_bR_c}{R_T}$$

Note the similarity between above Equation and $R_1 = \dfrac{R_bR_c}{R_T}$, $R_2 = \dfrac{R_aR_c}{R_T}$, $R_3 = \dfrac{R_aR_b}{R_T}$.

Divide above equation by $R_1 = \dfrac{R_bR_c}{R_T}$,

$$\frac{R_1R_2 + R_1R_3 + R_2R_3}{R_1} = \frac{R_aR_bR_c}{R_T}\frac{R_T}{R_bR_c}$$

$$= R_a,$$

which is the equation for R_a. Dividing $= \dfrac{R_aR_bR_c}{R_T}$ by $R_2 = \dfrac{R_aR_c}{R_T}$ or $R_3 = \dfrac{R_aR_b}{R_T}$. (expressions for R_2 or R_3) gives the remaining equations.

DELTA STAR TRANSFORMATION

Three branches in an electrical network can be connected in numbers of forms but most common among them is either star or delta form. In delta connection, three branches are so connected, that they form a closed loop. As these three branches are connected nose to tail, they form a triangular closed loop, this configuration is referred as delta connection. On the other hand, when either terminal of three branches is connected to a common point to form a Y like pattern is known as star connection. But these star and delta connections can be transformed from one form to another. For simplifying complex network, delta to star or star to delta transformation is often required.

Delta to Star Conversion

The replacement of delta or mesh by equivalent star connection is known as delta – star transformation. The two connections are equivalent or identical to each other if the impedance is measured between any pair of lines. That means, the value of impedance will be the same if it is measured between any pair of lines irrespective of whether the delta is connected between the lines or its equivalent star is connected between that lines.

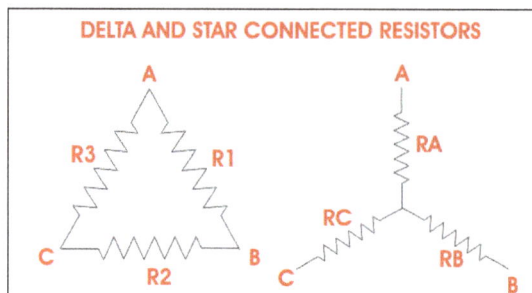

DELTA AND STAR CONNECTED RESISTORS

Consider a delta system that's three corner points are A, B and C as shown in the figure. Electrical resistance of the branch between points A and B, B and C and C and A are R_1, R_2 and R_3 respectively. The resistance between the points A and B will be,

$$R_{AB} = R_1 \| (R_2 + R_3) = \frac{R_1 \cdot (R_2 + R_3)}{R_1 + R_2 + R_3}$$

Now, one star system is connected to these points A, B, and C as shown in the figure. Three arms R_A, R_B and R_C of the star system are connected with A, B and C respectively. Now if we measure the resistance value between points A and B, we will get,

$$R_{AB} = R_A + R_B$$

Since the two systems are identical, resistance measured between terminals A and B in both systems must be equal.

$$R_A + R_B = \frac{R_1 \cdot (R_2 + R_3)}{R_1 + R_2 + R_3}$$

Similarly, resistance between points B and C being equal in the two systems,

$$R_B + R_C = \frac{R_2 \cdot (R_3 + R_1)}{R_1 + R_2 + R_3}$$

And resistance between points C and A being equal in the two systems,

$$R_C + R_A = \frac{R_3 \cdot (R_1 + R_2)}{R_1 + R_2 R_3}$$

Adding equations $R_A + R_B = \dfrac{R_1 \cdot (R_2 + R_3)}{R_1 + R_2 + R_3}$, $R_B + R_C = \dfrac{R_2 \cdot (R_3 + R_1)}{R_1 + R_2 + R_3}$ and $R_C + R_A = \dfrac{R_3 \cdot (R_1 + R_2)}{R_1 + R_2 R_3}$ we get,

$$2(R_A + R_B + R_C) = \frac{2(R_1 \cdot R_2 + R_2 \cdot R_3 + R_3 \cdot R_1)}{R_1 + R_2 + R_3}$$

$$R_A + R_B + R_C = \frac{R_1 \cdot R_2 + R_2 \cdot R_3 + R_3 \cdot R_1}{R_1 + R_2 + R_3}$$

Subtracting equations $R_A + R_B = \dfrac{R_1 \cdot (R_2 + R_3)}{R_1 + R_2 + R_3}$, $R_B + R_C = \dfrac{R_2 \cdot (R_3 + R_1)}{R_1 + R_2 + R_3}$ and $R_C + R_A = \dfrac{R_3 \cdot (R_1 + R_2)}{R_1 + R_2 R_3}$

from equation $R_A + R_B + R_C = \dfrac{R_1 \cdot R_2 + R_2 \cdot R_3 + R_3 \cdot R_1}{R_1 + R_2 + R_3}$ we get,

$$R_A = \frac{R_3 \cdot R_1}{R_1 + R_2 + R_3}$$

$$R_B = \frac{R_1.R_2}{R_1 + R_2 + R_3}$$

$$R_C = \frac{R_2.R_3}{R_1 + R_2 + R_3}$$

The relation of delta – star transformation can be expressed as follows.

The equivalent star resistance connected to a given terminal, is equal to the product of the two delta resistances connected to the same terminal divided by the sum of the delta connected resistances.

If the delta connected system has same resistance R at its three sides then equivalent star resistance r will be,

$$r = \frac{R.R}{R + R + R +} = \frac{R}{3}$$

Star to Delta Conversion

For star – delta transformation we just multiply equations $R_A = \dfrac{R_3.R_1}{R_1 + R_2 + R_3}$, $R_B = \dfrac{R_1.R_2}{R_1 + R_2 + R_3}$

and $R_B = \dfrac{R_1.R_2}{R_1 + R_2 + R_3}$, $R_C = \dfrac{R_2.R_3}{R_1 + R_2 + R_3}$ and $R_C = \dfrac{R_2.R_3}{R_1 + R_2 + R_3}$, $R_A = \dfrac{R_3.R_1}{R_1 + R_2 + R_3}$ that is

by doing $R_A = \dfrac{R_3.R_1}{R_1 + R_2 + R_3} \times R_B = \dfrac{R_1.R_2}{R_1 + R_2 + R_3} + R_B = \dfrac{R_1.R_2}{R_1 + R_2 + R_3} \times R_C = \dfrac{R_2.R_3}{R_1 + R_2 + R_3} +$

$R_C = \dfrac{R_2.R_3}{R_1 + R_2 + R_3} \times R_A = \dfrac{R_3.R_1}{R_1 + R_2 + R_3}$ we get,

$$R_A R_B + R_B R_C + R_C R_A = \frac{R_1.R_2^2.R_3 + R_1.R_2.R_3^2 + R_1^2.R_2.R_3}{(R_1 + R_2 + R_3)^2}$$

$$= \frac{R_1.R_2.R_3(R_3 + R_2 + R_3)}{(R_1 + R_2 + R_3)^2}$$

$$= \frac{R_1.R_2.R_3}{R_1 + R_2 + R_3}$$

Now dividing equation $= \dfrac{R_1.R_2.R_3}{R_1 + R_2 + R_3}$ by equations $R_A = \dfrac{R_3.R_1}{R_1 + R_2 + R_3}$, $R_B = \dfrac{R_1.R_2}{R_1 + R_2 + R_3}$ and

equations $R_C = \dfrac{R_2.R_3}{R_1 + R_2 + R_3}$ separately we get,

$$R_3 = \frac{R_A R_B + R_B R_C + R_C R_A}{R_A}$$

$$R_1 = \frac{R_A R_B + R_B R_C + R_C R_A}{R_B}$$

$$R_2 = \frac{R_A R_B + R_B R_C + R_C R_A}{R_C}$$

MILLMAN THEOREM

Millman's theorem was named after famous electrical engineering professor jacob millman who proposed the idea of this theorem. Millman's theorem acts as a very strong tool in case of simplifying the special type of complex electrical circuit. This theorem is nothing but a combination of thevenin's theorem and norton's theorem. It is very useful theorem to find out voltage across the load and current through the load. This theorem is also called as parallel generator theorem.

Millman's theorem is applicable to a circuit which may contain only voltage sources in parallel or a mixture of voltage and current sources connected in parallel.

Circuit Consisting Only Voltage Sources

Let us have a circuit as shown in figure.

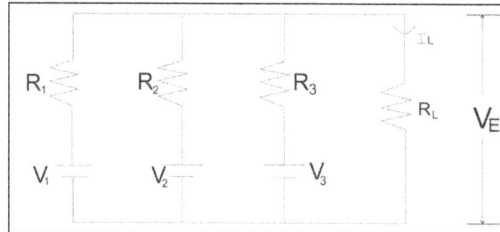

Here V_1, V_2 and V_3 are voltages of respectively 1st, 2nd and 3rd branch and R_1, R_2 and R_3 are their respective resistances. I_L, R_L and V_T are load current, load resistance and terminal voltage respectively. Now this complex circuit can be reduced easily to a single equivalent voltage source with a series resistance with the help of Millman's Theorem as shown in figure.

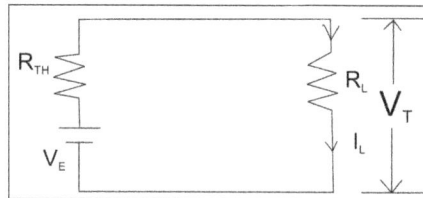

The value of equivalent voltage V_E is specified as per Millman's theorem will be,

$$V_E = \frac{\dfrac{V_1}{R_1} + \dfrac{V_2}{R_2} + \dfrac{V_3}{R_3}}{\dfrac{1}{R_1} + \dfrac{1}{R_2} + \dfrac{1}{R_3}} = \frac{\sum \dfrac{V}{R}}{\sum \dfrac{1}{R}}$$

This V_E is nothing but Thevenin voltage and Thevenin resistance R_{TH} can be determined as per convention by shorting the voltage source. So R_{TH} will be obtained as:

$$R_{TH} = \frac{1}{\dfrac{1}{R_1} + \dfrac{1}{R_2} + \dfrac{1}{R_3}}$$

Now load current and terminal voltage can be easily found by,

$$I_L = \frac{V_{TH}}{R_L + R_{TH}} \text{ and } V_T = I_L \times R_L$$

Let's try to understand whole concept of Millman's Theorem with the help of a example.

Example:

A circuit is given as shown in figure. Find out the voltage across 2 Ohm resistance and current through the 2 ohm resistance.

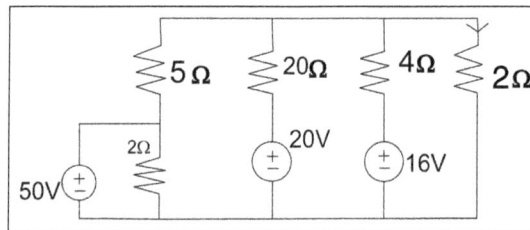

We can go through any solving method to solve this problem but the most effecting and time saving method will be none another than Millman's theorem. Given circuit can be reduced to a circuit shown in figure where equivalent voltage V_E can be obtained by millman's theorem and that is,

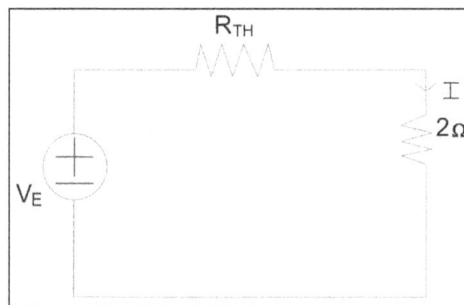

$$V_E = \frac{\dfrac{50}{5} - \dfrac{20}{20} + \dfrac{16}{4}}{\dfrac{1}{5} + \dfrac{1}{20} + \dfrac{1}{4}} = 26V$$

Equivalent resistance or Thevenin resistance can be found by shorting the voltage sources as shown in figure.

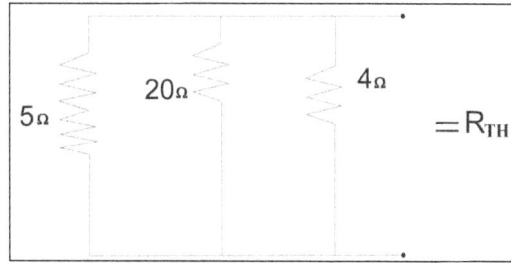

$$R_{TH} = \cfrac{1}{\cfrac{1}{5} + \cfrac{1}{20} + \cfrac{1}{4}} = 2\Omega$$

Now we can easily found the required current through 2 Ohm load resistance by Ohm's law.

$$I_{2\Omega} = \frac{26}{2+2} + 6.5\,A$$

Voltage across load is,

$$V_L = I_{2\Omega} \times 2 = 6.5 \times 2 = 13V$$

Circuit is Consisting Mixture of Voltage and Current Source

Millman's Theorem is also helpful to reduce a mixture of voltage and current source connected in parallel to a single equivalent voltage or current source. Let's have a circuit as shown in below figure.

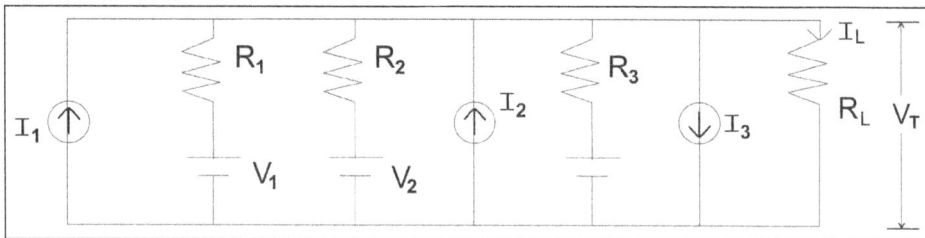

Here all letters are implying their conventional representation. This circuit can be reduced to a circuit as shown in figure.

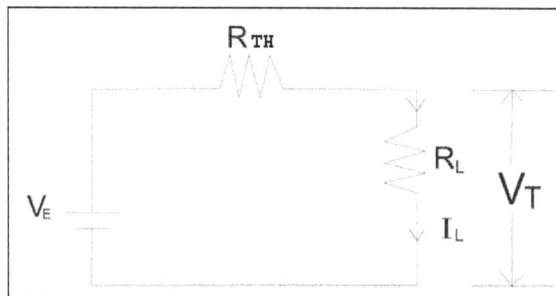

Here V_E which is nothing but thevenin voltage which will be obtained as per Millman's theorem

and that is,

$$V_E = \frac{\dfrac{V_1}{R_1} + \dfrac{V_2}{R_2} + \dfrac{V_3}{R_3} + I_1 + I_2 - I_3}{\dfrac{1}{R_1} + \dfrac{1}{R_2} + \dfrac{1}{R_3}} = \frac{\sum \dfrac{V}{R} + \sum I}{\sum \dfrac{1}{R}}$$

And R_{TH} will be obtained by replacing current sources with open circuits and voltage sources with short circuits.

$$R_{TH} = \frac{1}{\dfrac{1}{R_1} + \dfrac{1}{R_2} + \dfrac{1}{R_3}}$$

Now we can easily find out load current I_L and terminal voltage V_T by Ohm's law.

$$I_L = \frac{V_E}{R_L + R_{TH}} \quad \text{and} \quad V_T = I_L \times R_L$$

Let's have a example to understand this concept more properly.

Example:

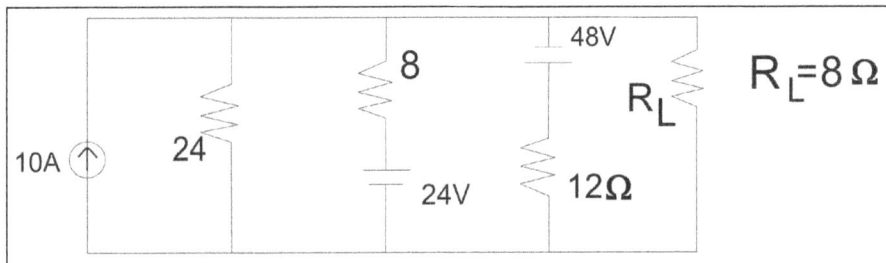

A circuit is given as shown in figure. Find out the current through load resistance where $R_L = 8\ \Omega$.

This problem may seem to be difficult to solve and time consuming but it can easily be solved in a very less time with the help of Millman's Theorem. The given circuit can be reduced in a circuit as shown in figure. Where, V_E can be obtained with the help of Millman's theorem,

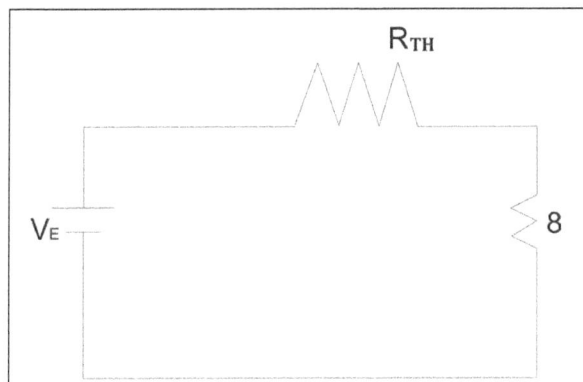

$$V_E = \frac{10 + \dfrac{0}{24} + \dfrac{24}{8} - \dfrac{48}{12}}{\dfrac{1}{24} + \dfrac{1}{8} + \dfrac{1}{12}} = 36V \text{ and } R_{TH} = 6\Omega$$

Therefore, current through load resistance 8 Ω is,

$$I_{8\Omega} = \frac{36}{6+8} = 2.57\ A.$$

References

- Kirchhoffs-voltage-law, dccircuits: electronics-tutorials.ws, Retrieved 4 June, Retrieved 14 July, 2019 2019

- Olivier Darrigol, Electrodynamics from Ampère to Einstein, p.70, Oxford University Press, 2000 ISBN 0-19-850594-9

- Kirchhoffs-current-law, dccircuits: electronics-tutorials.ws, Retrieved 3 February, Retrieved 14 July, 2019 2019

- Akers; M. Gassman & R. Smith (2006). Hydraulic Power System Analysis. New York: Taylor & Francis. Chapter 13. ISBN 978-0-8247-9956-4

- Joules-law: electrical4u.com, Retrieved 26 August, 2019

- Edward Hughes revised by John.K, Keith.B etal (2008) Electrical and Electronic Technology (10th ed.) Pearson ISBN 978-0-13-206011-0 page 75-77

- Reciprocity-theorem, theorems-and-laws, knowledge, resources: electrical-engineering-portal.com, Retrieved 13 March, 2019

- Thompson Phillips (2009-05-30), Dynamo-Electric Machinery; A Manual for Students of Electrotechnics, bibliobazaar, LLC, ISBN 978-1-110-35104-6

- What-is-compensation-theorem: circuitglobe.com, Retrieved 1 January, 2019

Electronic Systems

The physical interconnection of components that gathers diverse amounts of information together is referred to as an electronic system. Some of the major focus areas related to electronic systems are system modelling, system analysis, system theory, complex adaptive system, etc. The chapter closely examines these key concepts of electronic systems to provide an extensive understanding of the subject.

An Electronic System is a physical interconnection of components, or parts, that gathers various amounts of information together It does this with the aid of input devices such as sensors, that respond in some way to this information and then uses electrical energy in the form of an output action to control a physical process or perform some type of mathematical operation on the signal.

But electronic control systems can also be regarded as a process that transforms one signal into another so as to give the desired system response. Then we can say that a simple electronic system consists of an input, a process, and an output with the input variable to the system and the output variable from the system both being signals.

There are many ways to represent a system, for example: mathematically, descriptively, pictorially or schematically. Electronic systems are generally represented schematically as a series of inter-connected blocks and signals with each block having its own set of inputs and outputs.

As a result, even the most complex of electronic control systems can be represented by a combination of simple blocks, with each block containing or representing an individual component or complete sub-system. The representing of an electronic system or process control system as a number of interconnected blocks or boxes is known commonly as "block-diagram representation".

Block Diagram Representation of a Simple Electronic System

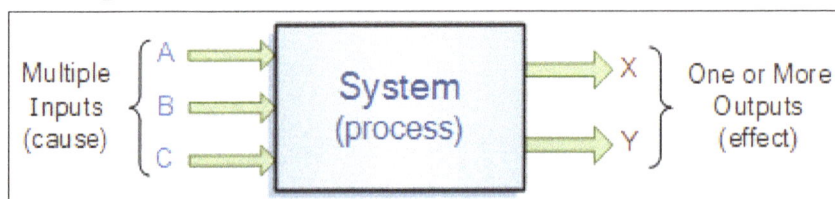

Electronic Systems have both Inputs and Outputs with the output or outputs being produced by processing the inputs. Also, the input signal(s) may cause the process to change or may itself cause the operation of the system to change. Therefore the input(s) to a system is the "cause" of the change, while the resulting action that occurs on the systems output due to this cause being present is called the "effect", with the effect being a consequence of the cause.

In other words, an electronic system can be classed as "causal" in nature as there is a direct relationship between its input and its output. Electronic systems analysis and process control theory are generally based upon this Cause and Effect analysis.

So for example in an audio system, a microphone (input device) causes sound waves to be converted into electrical signals for the amplifier to amplify (a process), and a loudspeaker (output device) produces sound waves as an effect of being driven by the amplifiers electrical signals.

But an electronic system need not be a simple or single operation. It can also be an interconnection of several sub-systems all working together within the same overall system.

Our audio system could for example, involve the connection of a CD player, or a DVD player, an MP3 player, or a radio receiver all being multiple inputs to the same amplifier which in turn drives one or more sets of stereo or home theatre type surround loudspeakers.

But an electronic system can not just be a collection of inputs and outputs, it must "do something", even if it is just to monitor a switch or to turn "ON" a light. We know that sensors are input devices that detect or turn real world measurements into electronic signals which can then be processed. These electrical signals can be in the form of either voltages or currents within a circuit. The opposite or output device is called an actuator, that converts the processed signal into some operation or action, usually in the form of mechanical movement.

Types of Electronic System

Electronic systems operate on either continuous-time (CT) signals or discrete-time (DT) signals. A continuous-time system is one in which the input signals are defined along a continuum of time, such as an analogue signal which "continues" over time producing a continuous-time signal.

But a continuous-time signal can also vary in magnitude or be periodic in nature with a time period T. As a result, continuous-time electronic systems tend to be purely analogue systems producing a linear operation with both their input and output signals referenced over a set period of time.

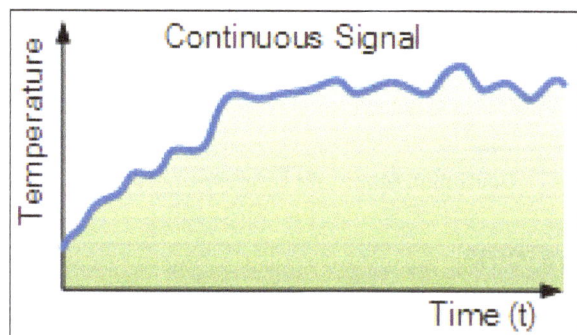

For example, the temperature of a room can be classed as a continuous time signal which can be measured between two values or set points, for example from cold to hot or from Monday to Friday. We can represent a continuous-time signal by using the independent variable for time t, and where $x(t)$ represents the input signal and $y(t)$ represents the output signal over a period of time t.

Generally, most of the signals present in the physical world which we can use tend to be continuous-time signals. For example, voltage, current, temperature, pressure, velocity, etc.

On the other hand, a discrete-time system is one in which the input signals are not continuous but a sequence or a series of signal values defined in "discrete" points of time. This results in a discrete-time output generally represented as a sequence of values or numbers.

Generally a discrete signal is specified only at discrete intervals, values or equally spaced points in time. So for example, the temperature of a room measured at 1pm, at 2pm, at 3pm and again at 4pm without regards for the actual room temperature in between these points at say, 1:30pm or at 2:45pm.

However, a continuous-time signal, x(t) can be represented as a discrete set of signals only at discrete intervals or "moments in time". Discrete signals are not measured versus time, but instead are plotted at discrete time intervals, where n is the sampling interval. As a result discrete-time signals are usually denoted as x(n) representing the input and y(n) representing the output.

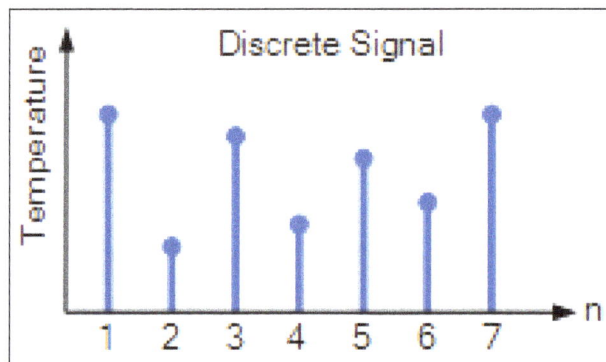

Then we can represent the input and output signals of a system as x and y respectively with the signal, or signals themselves being represented by the variable, t, which usually represents time for a continuous system and the variable n, which represents an integer value for a discrete system as shown.

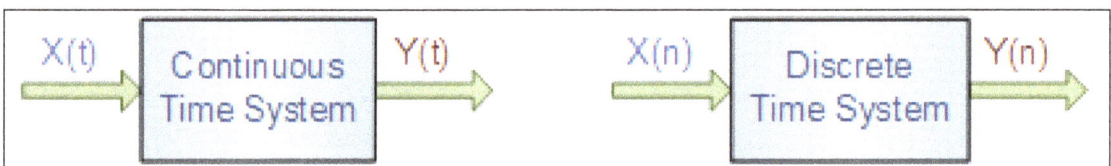

Continuous-time and Discrete-time System.

Interconnection of Systems

One of the practical aspects of electronic systems and block-diagram representation is that they can be combined together in either a series or parallel combinations to form much bigger systems. Many larger real systems are built using the interconnection of several sub-systems and by using block diagrams to represent each subsystem, we can build a graphical representation of the whole system being analysed.

When subsystems are combined to form a series circuit, the overall output at y(t) will be equivalent to the multiplication of the input signal x(t) as shown as the subsystems are cascaded together.

Series Connected System

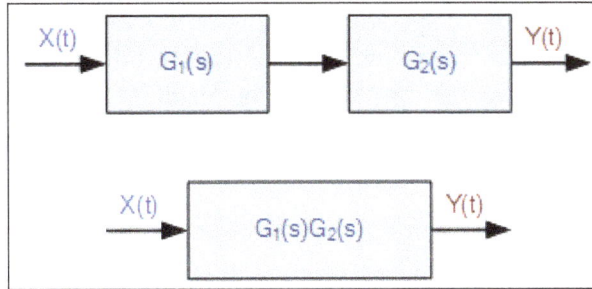

For a series connected continuous-time system, the output signal y(t) of the first subsystem, "A" becomes the input signal of the second subsystem, "B" whose output becomes the input of the third subsystem, "C" and so on through the series chain giving A x B x C, etc.

Then the original input signal is cascaded through a series connected system, so for two series connected subsystems, the equivalent single output will be equal to the multiplication of the systems, ie, $y(t) = G_1(s) \times G_2(s)$. Where G represents the transfer function of the subsystem.

Note that the term "Transfer Function" of a system refers to and is defined as being the mathematical relationship between the systems input and its output, or output/input and hence describes the behaviour of the system.

Also, for a series connected system, the order in which a series operation is performed does not matter with regards to the input and output signals as: $G_1(s) \times G_2(s)$ is the same as $G_2(s) \times G_1(s)$. An example of a simple series connected circuit could be a single microphone feeding an amplifier followed by a speaker.

Parallel Connected Electronic System

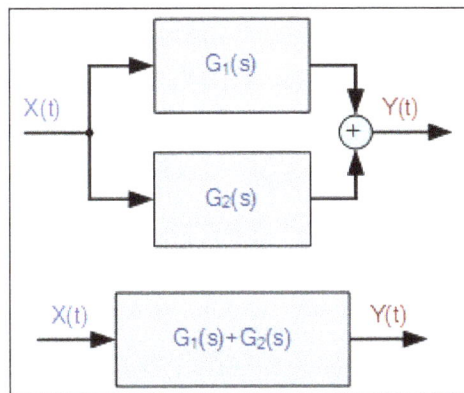

For a parallel connected continuous-time system, each subsystem receives the same input signal, and their individual outputs are summed together to produce an overall output, y(t). Then for two parallel connected subsystems, the equivalent single output will be the sum of the two individual inputs, ie, $y(t) = G_1(s) + G_2(s)$.

An example of a simple parallel connected circuit could be several microphones feeding into a mixing desk which in turn feeds an amplifier and speaker system.

Electronic Feedback Systems

Another important interconnection of systems which is used extensively in control systems, is the "feedback configuration". In feedback systems, a fraction of the output signal is "fed back" and either added to or subtracted from the original input signal. The result is that the output of the system is continually altering or updating its input with the purpose of modifying the response of a system to improve stability. A feedback system is also commonly referred to as a "Closed-loop System" as shown.

Closed-Loop Feedback System

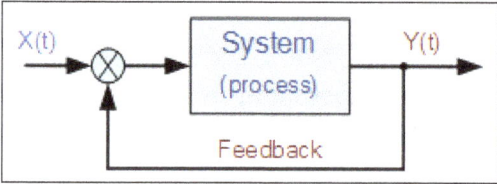

Feedback systems are used a lot in most practical electronic system designs to help stabilise the system and to increase its control. If the feedback loop reduces the value of the original signal, the feedback loop is known as "negative feedback". If the feedback loop adds to the value of the original signal, the feedback loop is known as "positive feedback".

An example of a simple feedback system could be a thermostatically controlled heating system in the home. If the home is too hot, the feedback loop will switch "OFF" the heating system to make it cooler. If the home is too cold, the feedback loop will switch "ON" the heating system to make it warmer. In this instance, the system comprises of the heating system, the air temperature and the thermostatically controlled feedback loop.

Transfer Function of Systems

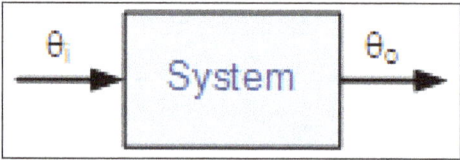

Any subsystem can be represented as a simple block with an input and output as shown. Generally, the input is designated as: θi and the output as: θo. The ratio of output over input represents the gain, (G) of the subsystem and is therefore defined as: $G = \theta o/\theta i$.

In this case, G represents the Transfer Function of the system or subsystem. When discussing electronic systems in terms of their transfer function, the complex operator, s is used, then the equation for the gain is rewritten as: $G(s) = \theta o(s)/\theta i(s)$.

Electronic System Summary

We have seen that a simple Electronic System consists of an input, a process, an output and possibly feedback. Electronic systems can be represented using interconnected block diagrams where the lines between each block or subsystem represents both the flow and direction of a signal through the system.

Block diagrams need not represent a simple single system but can represent very complex systems made from many interconnected subsystems. These subsystems can be connected together in series, parallel or combinations of both depending upon the flow of the signals.

We have also seen that electronic signals and systems can be of continuous-time or discrete-time in nature and may be analogue, digital or both. Feedback loops can be used be used to increase or reduce the performance of a particular system by providing better stability and control. Control is the process of making a system variable adhere to a particular value, called the reference value.

SYSTEM MODELING

System modeling is the process of developing abstract models of a system, with each model presenting a different view or perspective of that system. It is about representing a system using some kind of graphical notation, which is now almost always based on notations in the Unified Modeling Language (UML). Models help the analyst to understand the functionality of the system; they are used to communicate with customers.

Models can explain the system from different perspectives:

- An external perspective, where you model the context or environment of the system.

- An interaction perspective, where you model the interactions between a system and its environment, or between the components of a system.

- A structural perspective, where you model the organization of a system or the structure of the data that is processed by the system.

- A behavioral perspective, where you model the dynamic behavior of the system and how it responds to events.

Five types of UML diagrams that are the most useful for system modeling:

- Activity diagrams, which show the activities involved in a process or in data processing.

- Use case diagrams, which show the interactions between a system and its environment.

- Sequence diagrams, which show interactions between actors and the system and between system components.

- Class diagrams, which show the object classes in the system and the associations between these classes.

- State diagrams, which show how the system reacts to internal and external events.

Models of both new and existing system are used during requirements engineering. Models of the existing systems help clarify what the existing system does and can be used as a basis for discussing its strengths and weaknesses. These then lead to requirements for the new system. Models of the new system are used during requirements engineering to help explain the proposed requirements to other system stakeholders. Engineers use these models to discuss design proposals and to document the system for implementation.

Context and Process Models

Context models are used to illustrate the operational context of a system - they show what lies outside the system boundaries. Social and organizational concerns may affect the decision on where to position system boundaries. Architectural models show the system and its relationship with other systems.

System boundaries are established to define what is inside and what is outside the system. They show other systems that are used or depend on the system being developed. The position of the system boundary has a profound effect on the system requirements. Defining a system boundary is a political judgment since there may be pressures to develop system boundaries that increase/decrease the influence or workload of different parts of an organization.

Context models simply show the other systems in the environment, not how the system being developed is used in that environment. Process models reveal how the system being developed is used in broader business processes. UML activity diagrams may be used to define business process models.

The example below shows a UML activity diagram describing the process of involuntary detention and the role of MHC-PMS (mental healthcare patient management system) in it.

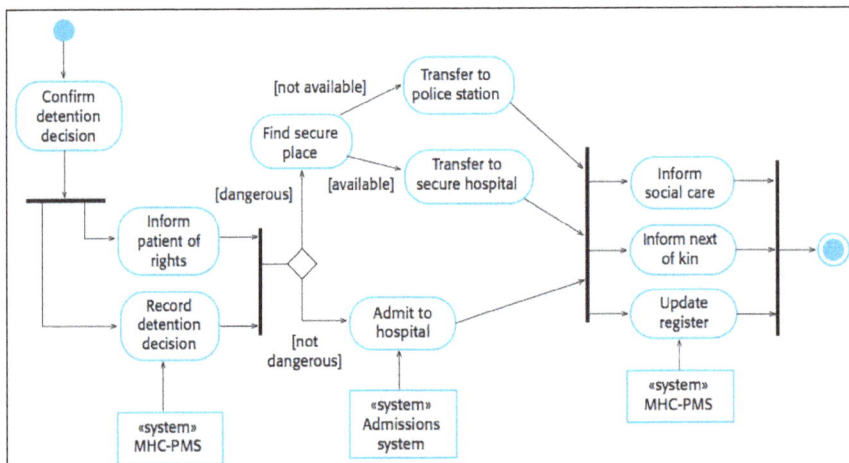

Interaction Models

Types of interactions that can be represented in a model:

- Modeling user interaction is important as it helps to identify user requirements.

- Modeling system-to-system interaction highlights the communication problems that may arise.

- Modeling component interaction helps us understand if a proposed system structure is likely to deliver the required system performance and dependability.

Use cases were developed originally to support requirements elicitation and now incorporated into the UML. Each use case represents a discrete task that involves external interaction with a system.

Actors in a use case may be people or other systems. Use cases can be represented using a UML use case diagram and in a more detailed textual/tabular format.

Simple use case diagram:

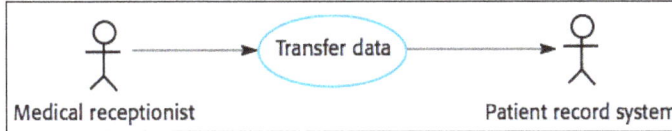

Use case description in a tabular format:

Use case title	Transfer data
Description	A receptionist may transfer data from the MHC-PMS to a general patient record database that is maintained by a health authority. The information transferred may either be updated personal information (address, phone number, etc.) or a summary of the patient's diagnosis and treatment.
Actor(s)	Medical receptionist, patient records system (PRS)
Preconditions	Patient data has been collected (personal information, treatment summary); The receptionist must have appropriate security permissions to access the patient information and the PRS.
Postconditions	PRS has been updated
Main success scenario	1. Receptionist selects the "Transfer data" option from the menu. 2. PRS verifies the security credentials of the receptionist. 3. Data is transferred. 4. PRS has been updated.
Extensions	2a. The receptionist does not have the necessary security credentials. 2a.1. An error message is displayed. 2a.2. The receptionist backs out of the use case.

UML sequence diagrams are used to model the interactions between the actors and the objects within a system. A sequence diagram shows the sequence of interactions that take place during a particular use case or use case instance. The objects and actors involved are listed along the top of the diagram, with a dotted line drawn vertically from these. Interactions between objects are indicated by annotated arrows.

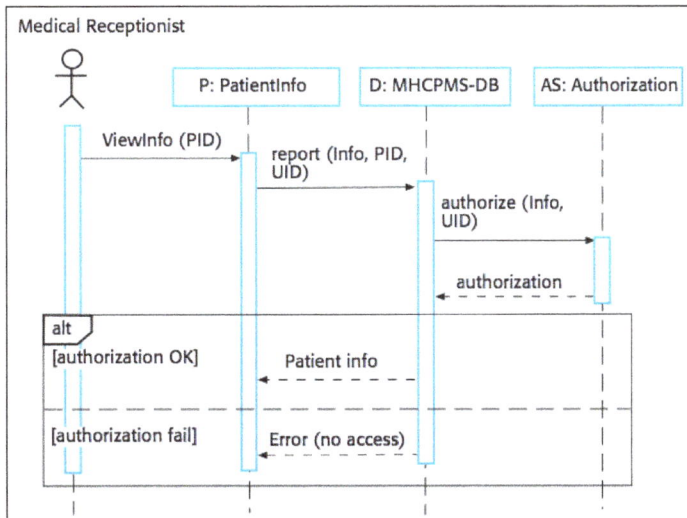

Structural Models

Structural models of software display the organization of a system in terms of the components that make up that system and their relationships. Structural models may be static models, which show the structure of the system design, or dynamic models, which show the organization of the system when it is executing. You create structural models of a system when you are discussing and designing the system architecture.

UML class diagrams are used when developing an object-oriented system model to show the classes in a system and the associations between these classes. An object class can be thought of as a general definition of one kind of system object. An association is a link between classes that indicates that there is some relationship between these classes. When you are developing models during the early stages of the software engineering process, objects represent something in the real world, such as a patient, a prescription, doctor, etc.

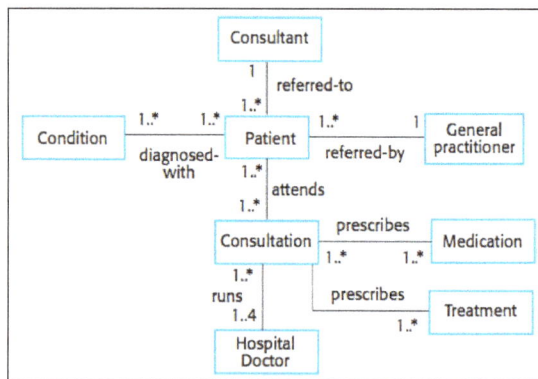

Generalization is an everyday technique that we use to manage complexity. In modeling systems, it is often useful to examine the classes in a system to see if there is scope for generalization. In object-oriented languages, such as Java, generalization is implemented using the class inheritance mechanisms built into the language. In a generalization, the attributes and operations associated with higher-level classes are also associated with the lower-level classes. The lower-level classes are subclasses inherit the attributes and operations from their superclasses. These lower-level classes then add more specific attributes and operations.

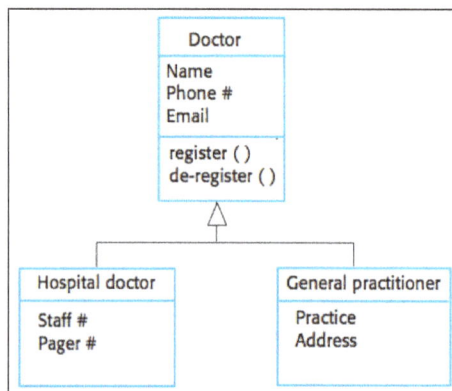

An aggregation model shows how classes that are collections are composed of other classes. Aggregation models are similar to the part-of relationship in semantic data models.

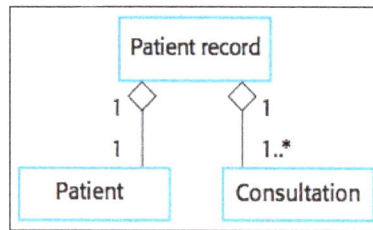

Behavioral Models

Behavioral models are models of the dynamic behavior of a system as it is executing. They show what happens or what is supposed to happen when a system responds to a stimulus from its environment. Two types of stimuli:

- Some data arrives that has to be processed by the system.

- Some event happens that triggers system processing. Events may have associated data, although this is not always the case.

Many business systems are data-processing systems that are primarily driven by data. They are controlled by the data input to the system, with relatively little external event processing. Data-driven models show the sequence of actions involved in processing input data and generating an associated output. They are particularly useful during the analysis of requirements as they can be used to show end-to-end processing in a system. Data-driven models can be created using UML activity diagrams:

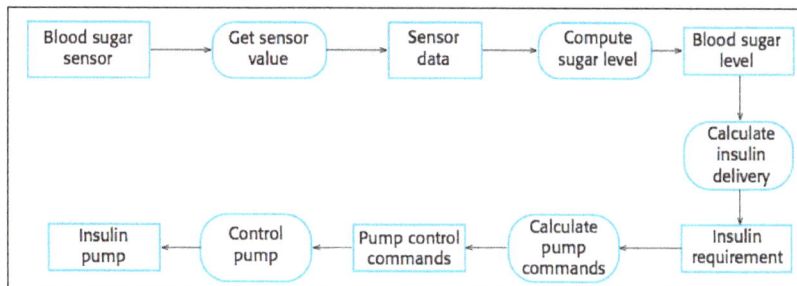

Data-driven models can also be created using UML sequence diagrams:

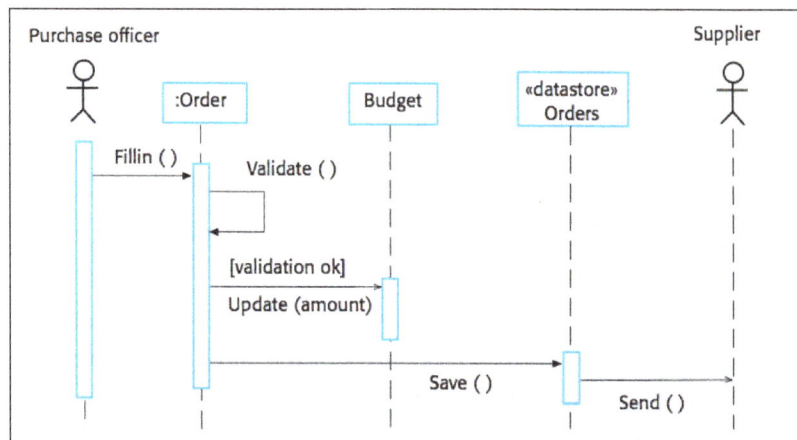

Real-time systems are often event-driven, with minimal data processing. For example, a landline phone switching system responds to events such as 'receiver off hook' by generating a dial tone. Event-driven models shows how a system responds to external and internal events. It is based on the assumption that a system has a finite number of states and that events (stimuli) may cause a transition from one state to another. Event-driven models can be created using UML state diagrams:

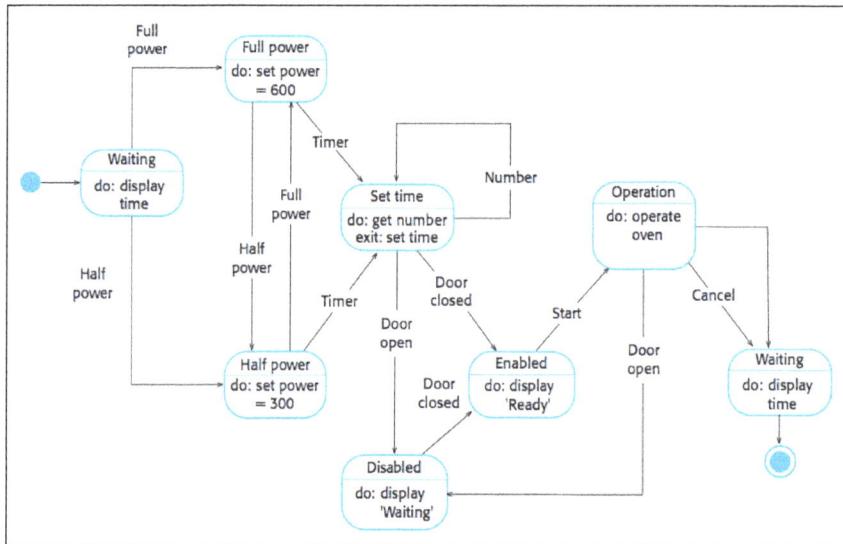

SYSTEM ANALYSIS

System analysis is the process of studying a procedure or business in order to identify its goals and purposes and create systems and procedures that will achieve them in an efficient way". Another view sees system analysis as a problem-solving technique that breaks down a system into its component pieces for the purpose of the studying how well those component parts work and interact to accomplish their purpose.

The field of system analysis relates closely to requirements analysis or to operations research. It is also "an explicit formal inquiry carried out to help a decision maker identify a better course of action and make a better decision than she might otherwise have made."

These terms are used in many scientific disciplines, from mathematics and logic to economics and psychology, to denote similar investigative procedures. Analysis is defined as "the procedure by which we break down an intellectual or substantial whole into parts," while synthesis means "the procedure by which we combine separate elements or components in order to form a coherent whole." System analysis researchers apply methodology to the systems involved, forming an overall picture.

System analysis is used in every field where something is developed. Analysis can also be a series of components that perform organic functions together, such as system engineering. System engineering is an interdisciplinary field of engineering that focuses on how complex engineering projects should be designed and managed.

Information Technology

The development of a computer-based information system includes a system analysis phase. This helps produce the data model, a precursor to creating or enhancing a database. There are a number of different approaches to system analysis. When a computer-based information system is developed, system analysis (according to the Waterfall model) would constitute the following steps:

- The development of a feasibility study: determining whether a project is economically, socially, technologically and organizationally feasible.

- Fact-finding measures, designed to ascertain the requirements of the system's end-users (typically involving interviews, questionnaires, or visual observations of work on the existing system).

- Gauging how the end-users would operate the system (in terms of general experience in using computer hardware or software), what the system would be used for and so on.

Another view outlines a phased approach to the process. This approach breaks system analysis into 5 phases:

- Scope Definition: Clearly defined objectives and requirements necessary to meet a project's requirements as defined by its stakeholders.

- Problem analysis: the process of understanding problems and needs and arriving at solutions that meet them.

- Requirements analysis: determining the conditions that need to be met.

- Logical design: looking at the logical relationship among the objects.

- Decision analysis: making a final decision.

Use cases are widely used system analysis modeling tools for identifying and expressing the functional requirements of a system. Each use case is a business scenario or event for which the system must provide a defined response. Use cases evolved from object-oriented analysis.

Practitioners

Practitioners of system analysis are often called up to dissect systems that have grown haphazardly to determine the current components of the system. This was shown during the year 2000 re-engineering effort as business and manufacturing processes were examined as part of the Y2K automation upgrades. Employment utilizing system analysis include system analyst, business analyst, manufacturing engineer, system architect, enterprise architect, software architect, etc.

While practitioners of system analysis can be called upon to create new systems, they often modify, expand or document existing systems (processes, procedures and methods). Researchers and practitioners rely on system analysis. Activity system analysis has been already applied to various research and practice studies including business management, educational reform, educational technology, etc.

SYSTEMS THEORY

Systems theory is the interdisciplinary study of systems. A system is a cohesive conglomeration of interrelated and interdependent parts that is either natural or man-made. Every system is delineated by its spatial and temporal boundaries, surrounded and influenced by its environment, described by its structure and purpose or nature and expressed in its functioning. In terms of its effects, a system can be more than the sum of its parts if it expresses synergy or emergent behavior. Changing one part of the system usually affects other parts and the whole system, with predictable patterns of behavior. For systems that are self-learning and self-adapting, the positive growth and adaptation depend upon how well the system is adjusted with its environment. Some systems function mainly to support other systems by aiding in the maintenance of the other system to prevent failure. The goal of systems theory is systematically discovering a system's dynamics, constraints, conditions and elucidating principles (purpose, measure, methods, tools, etc.) that can be discerned and applied to systems at every level of nesting, and in every field for achieving optimized equifinality.

General systems theory is about broadly applicable concepts and principles, as opposed to concepts and principles applicable to one domain of knowledge. It distinguishes dynamic or active systems from static or passive systems. Active systems are activity structures or components that interact in behaviours and processes. Passive systems are structures and components that are being processed. E.g. a program is passive when it is a disc file and active when it runs in memory. The field is related to systems thinking, machine logic and systems engineering.

Key Concepts

- System: An organized entity made up of interrelated and interdependent parts.

- Boundaries: Barriers that define a system and distinguish it from other systems in the environment.

- Homeostasis: The tendency of a system to be resilient towards external factors and maintain its key characteristics.

- Adaptation: The tendency of a self-adapting system to make the internal changes needed to protect itself and keep fulfilling its purpose.

- Reciprocal Transactions: Circular or cyclical interactions that systems engage in such that they influence one another.

- Feedback Loop: The process by which systems self-correct based on reactions from other systems in the environment.

- Throughput: Rate of energy transfer between the system and its environment during the time it is functioning.

- Microsystem: The system closest to the client.

- Mesosystem: Relationships among the systems in an environment.

- Exosystem: A relationship between two systems that has an indirect effect on a third system.

- Macrosystem: A larger system that influences clients, such as policies, administration of entitlement programs, and culture.

- Chronosystem: A system composed of significant life events that can affect adaptation.

Contemporary ideas from systems theory have grown with diverse areas, exemplified by the work of biologist Ludwig von Bertalanffy, linguist Béla H. Bánáthy, sociologist Talcott Parsons, ecological systems with Howard T. Odum, Eugene Odum and Fritjof Capra, organizational theory and management with individuals such as Peter Senge, interdisciplinary study with areas like Human Resource Development from the work of Richard A. Swanson, and insights from educators such as Debora Hammond and Alfonso Montuori. As a transdisciplinary, interdisciplinary and multiperspectival domain, the area brings together principles and concepts from ontology, philosophy of science, physics, computer science, biology and engineering as well as geography, sociology, political science, psychotherapy (within family systems therapy) and economics among others. Systems theory thus serves as a bridge for interdisciplinary dialogue between autonomous areas of study as well as within the area of systems science itself.

In this respect, with the possibility of misinterpretations, von Bertalanffy believed a general theory of systems "should be an important regulative device in science", to guard against superficial analogies that "are useless in science and harmful in their practical consequences". Others remain closer to the direct systems concepts developed by the original theorists. For example, Ilya Prigogine, of the Center for Complex Quantum Systems at the University of Texas, Austin, has studied emergent properties, suggesting that they offer analogues for living systems. The theories of autopoiesis of Francisco Varela and Humberto Maturana represent further developments in this field. Important names in contemporary systems science include Russell Ackoff, Ruzena Bajcsy, Béla H. Bánáthy, Gregory Bateson, Anthony Stafford Beer, Peter Checkland, Barbara Grosz, Brian Wilson, Robert L. Flood, Allenna Leonard, Radhika Nagpal, Fritjof Capra, Warren McCulloch, Kathleen Carley, Michael C. Jackson, Katia Sycara, and Edgar Morin among others.

With the modern foundations for a general theory of systems following World War I, Ervin Laszlo, in the preface for Bertalanffy's book: *Perspectives on General System Theory*, points out that the translation of "general system theory" from German into English has "wrought a certain amount of havoc":

> It (General System Theory) was criticized as pseudoscience and said to be nothing more than an admonishment to attend to things in a holistic way. Such criticisms would have lost their point had it been recognized that von Bertalanffy's general system theory is a perspective or paradigm, and that such basic conceptual frameworks play a key role in the development of exact scientific theory. Allgemeine Systemtheorie is not directly consistent with an interpretation often put on 'general system theory,' to wit, that it is a (scientific) "theory of general systems." To criticize it as such is to shoot at straw men. Von Bertalanffy opened up something much broader and of much greater significance than a single theory (which, as we now know, can always be falsified and has usually an ephemeral existence): he created a new paradigm for the development of theories.

"Theorie" (or "Lehre"), just as "Wissenschaft" (translated Scholarship), "has a much broader meaning in German than the closest English words 'theory' and 'science'". These ideas refer to an organized body of knowledge and "any systematically presented set of concepts, whether empirically, axiomatically, or philosophically" represented, while many associate "Lehre" with theory and science in the etymology of general systems, though it also does not translate from the German very well; its "closest equivalent" translates as "teaching", but "sounds dogmatic and off the mark". While the idea of a "general systems theory" might have lost many of its root meanings in the translation, by defining a new way of thinking about science and scientific paradigms, Systems theory became a widespread term used for instance to describe the interdependence of relationships created in organizations.

A system in this frame of reference can contain regularly interacting or interrelating groups of activities. For example, in noting the influence in organizational psychology as the field evolved from "an individually oriented industrial psychology to a systems and developmentally oriented organizational psychology", some theorists recognize that organizations have complex social systems; separating the parts from the whole reduces the overall effectiveness of organizations. This difference, from conventional models that center on individuals, structures, departments and units, separates in part from the whole, instead of recognizing the interdependence between groups of individuals, structures and processes that enable an organization to function. Laszlo explains that the new systems view of organized complexity went "one step beyond the Newtonian view of organized simplicity" which reduced the parts from the whole, or understood the whole without relation to the parts. The relationship between organisations and their environments can be seen as the foremost source of complexity and interdependence. In most cases, the whole has properties that cannot be known from analysis of the constituent elements in isolation. Béla H. Bánáthy, who argued—along with the founders of the systems society—that "the benefit of humankind" is the purpose of science, has made significant and far-reaching contributions to the area of systems theory. For the Primer Group at ISSS, Bánáthy defines a perspective that iterates this view:

> The systems view is a world-view that is based on the discipline of System Inquiry. Central to systems inquiry is the concept of System. In the most general sense, system means a configuration of parts connected and joined together by a web of relationships. The Primer Group defines system as a family of relationships among the members acting as a whole. Von Bertalanffy defined system as "elements in standing relationship."

Similar ideas are found in learning theories that developed from the same fundamental concepts, emphasising how understanding results from knowing concepts both in part and as a whole. In fact, Bertalanffy's organismic psychology paralleled the learning theory of Jean Piaget. Some consider interdisciplinary perspectives critical in breaking away from industrial age models and thinking, wherein history represents history and math represents math, while the arts and sciences specialization remain separate and many treat teaching as behaviorist conditioning. The contemporary work of Peter Senge provides detailed discussion of the commonplace critique of educational systems grounded in conventional assumptions about learning, including the problems with fragmented knowledge and lack of holistic learning from the "machine-age thinking" that became a "model of school separated from daily life". In this way some systems theorists attempt to provide alternatives to, and evolved ideation from orthodox theories which have grounds in classical assumptions, including individuals such as Max Weber and Émile Durkheim in sociology and Frederick Winslow Taylor in scientific management. The theorists sought holistic methods by developing systems concepts that could integrate with different areas.

Some may view the contradiction of reductionism in conventional theory (which has as its subject a single part) as simply an example of changing assumptions. The emphasis with systems theory shifts from parts to the organization of parts, recognizing interactions of the parts as not static and constant but dynamic processes. Some questioned the conventional closed systems with the development of open systems perspectives. The shift originated from absolute and universal authoritative principles and knowledge to relative and general conceptual and perceptual knowledge and still remains in the tradition of theorists that sought to provide means to organize human life. In other words, theorists rethought the preceding history of ideas; they did not lose them. Mechanistic thinking was particularly critiqued, especially the industrial-age mechanistic metaphor for the mind from interpretations of Newtonian mechanics by Enlightenment philosophers and later psychologists that laid the foundations of modern organizational theory and management by the late 19th century.

Developments

General Systems Research and Systems Inquiry

Many early systems theorists aimed at finding a general systems theory that could explain all systems in all fields of science. The term goes back to Bertalanffy's book titled *"General System theory: Foundations, Development, Applications"* from 1968. He developed the "allgemeine Systemlehre" (general systems theory) first via lectures beginning in 1937 and then via publications beginning in 1946.

Von Bertalanffy's objective was to bring together under one heading the organismic science he had observed in his work as a biologist. His desire was to use the word *system* for those principles that are common to systems in general. In GST, he writes:

> "There exist models, principles, and laws that apply to generalized systems or their subclasses, irrespective of their particular kind, the nature of their component elements, and the relationships or "forces" between them. It seems legitimate to ask for a theory, not of systems of a more or less special kind, but of universal principles applying to systems in general".

> *— Von Bertalanffy*

Ervin Laszlo in the preface of von Bertalanffy's book *Perspectives on General System Theory*:

> Thus when von Bertalanffy spoke of Allgemeine Systemtheorie it was consistent with his view that he was proposing a new perspective, a new way of doing science. It was not directly consistent with an interpretation often put on "general system theory", to wit, that it is a (scientific) "theory of general systems." To criticize it as such is to shoot at straw men. Von Bertalanffy opened up something much broader and of much greater significance than a single theory (which, as we now know, can always be falsified and has usually an ephemeral existence): he created a new paradigm for the development of theories.

Ludwig von Bertalanffy outlines systems inquiry into three major domains: Philosophy, Science, and Technology. In his work with the Primer Group, Béla H. Bánáthy generalized the domains into four integratable domains of systemic inquiry:

Domain	Description
Philosophy	The ontology, epistemology and axiology of systems
Theory	a set of interrelated concepts and principles applying to all systems
Methodology	The set of models, strategies, methods and tools that instrumentalize systems theory and philosophy
Application	The application and interaction of the domains

These operate in a recursive relationship, he explained. Integrating Philosophy and Theory as Knowledge, and Method and Application as action, Systems Inquiry then is knowledgeable action.

Cybernetics

Cybernetics is the study of the communication and control of regulatory feedback both in living and lifeless systems (organisms, organizations, machines), and in combinations of those. Its focus is how anything (digital, mechanical or biological) controls its behavior, processes information, reacts to information, and changes or can be changed to better accomplish those three primary tasks.

The terms "systems theory" and "cybernetics" have been widely used as synonyms. Some authors use the term *cybernetic* systems to denote a proper subset of the class of general systems, namely those systems that include feedback loops. However Gordon Pask's differences of eternal interacting actor loops (that produce finite products) makes general systems a proper subset of cybernetics. According to Jackson, von Bertalanffy promoted an embryonic form of general system theory (GST) as early as the 1920s and 1930s but it was not until the early 1950s it became more widely known in scientific circles.

Threads of cybernetics began in the late 1800s that led toward the publishing of seminal works (e.g., Wiener's *Cybernetics* in 1948 and von Bertalanffy's *General Systems Theory* in 1968). Cybernetics arose more from engineering fields and GST from biology. If anything it appears that although the two probably mutually influenced each other, cybernetics had the greater influence. Von Bertalanffy specifically makes the point of distinguishing between the areas in noting the influence of cybernetics: "Systems theory is frequently identified with cybernetics and control theory. This again is incorrect. Cybernetics as the theory of control mechanisms in technology and nature is founded on the concepts of information and feedback, but as part of a general theory of systems;" then reiterates: "the model is of wide application but should not be identified with 'systems theory' in general", and that "warning is necessary against its incautious expansion to fields for which its concepts are not made." (17-23). Jackson also claims von Bertalanffy was informed by Alexander Bogdanov's three volume *Tectology* that was published in Russia between 1912 and 1917, and was translated into German in 1928. He also states it is clear to Gorelik that the "conceptual part" of general system theory (GST) had first been put in place by Bogdanov. The similar position is held by Mattessich and Capra. Ludwig von Bertalanffy never even mentioned Bogdanov in his works, which Capra finds "surprising".

Cybernetics, catastrophe theory, chaos theory and complexity theory have the common goal to explain complex systems that consist of a large number of mutually interacting and interrelated parts in terms of those interactions. Cellular automata (CA), neural networks (NN), artificial intelligence (AI), and artificial life (ALife) are related fields, but they do not try to describe general (universal) complex (singular) systems. The best context to compare the different "C"-Theories about complex systems is historical, which emphasizes different tools and methodologies, from pure mathematics in the beginning to pure computer science now. Since the beginning of chaos theory when Edward Lorenz accidentally discovered a strange attractor with his computer, computers have become an indispensable source of information. One could not imagine the study of complex systems without the use of computers today.

Complex Adaptive Systems

Complex adaptive systems (CAS) are special cases of complex systems. They are *complex* in that they are diverse and composed of multiple, interconnected elements; they are *adaptive* in that they have the capacity to change and learn from experience. In contrast to control systems in which negative feedback dampens and reverses disequilibria, CAS are often subject to positive feedback, which magnifies and perpetuates changes, converting local irregularities into global features. Another mechanism, Dual-phase evolution arises when connections between elements repeatedly change, shifting the system between phases of variation and selection that reshape the system. Differently from Beer Management Cybernetics, Cultural Agency Theory (CAT) provides a modelling approach to explore predefined contexts and can be adapted to reflect those contexts.

The term *complex adaptive system* was coined at the interdisciplinary Santa Fe Institute (SFI), by John H. Holland, Murray Gell-Mann and others. An alternative conception of complex adaptive (and learning) systems, methodologically at the interface between natural and social science, has been presented by Kristo Ivanov in terms of hypersystems. This concept intends to offer a theoretical basis for understanding and implementing participation of "users", decisions makers, designers and affected actors, in the development or maintenance of self-learning systems.

COMPLEX SYSTEM

A complex system is a system composed of many components which may interact with each other. Examples of complex systems are Earth's global climate, organisms, the human brain, infrastructure such as power grid, transportation or communication systems, social and economic organizations (like cities), an ecosystem, a living cell, and ultimately the entire universe.

Complex systems are systems whose behavior is intrinsically difficult to model due to the dependencies, competitions, relationships, or other types of interactions between their parts or between a given system and its environment. Systems that are "complex" have distinct properties that arise from these relationships, such as nonlinearity, emergence, spontaneous order, adaptation, and feedback loops, among others. Because such systems appear in a wide variety of fields, the commonalities among them have become the topic of their own independent area of research. In many cases it is useful to represent such a system as a network where the nodes represent the components and the links their interactions.

The term *complex systems* often refers to the study of complex systems, which is an approach to science that investigates how relationships between a system's parts give rise to its collective behaviors and how the system interacts and forms relationships with its environment. The study of complex systems regards collective, or system-wide, behaviors as the fundamental object of study; for this reason, complex systems can be understood as an alternative paradigm to reductionism, which attempts to explain systems in terms of their constituent parts and the individual interactions between them.

As an interdisciplinary domain, complex systems draws contributions from many different fields, such as the study of self-organization from physics, that of spontaneous order from the social sciences, chaos from mathematics, adaptation from biology, and many others. *Complex systems* is therefore often used as a broad term encompassing a research approach to problems in many diverse disciplines, including statistical physics, information theory, nonlinear dynamics, anthropology, computer science, meteorology, sociology, economics, psychology, and biology.

Key Concepts

Systems

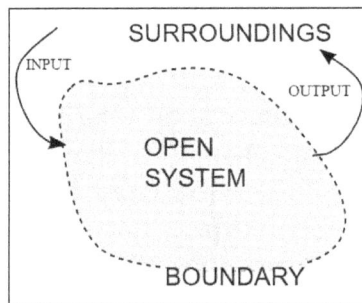

Open systems have input and output flows, representing
exchanges of matter, energy or information with their surroundings.

Complex systems is chiefly concerned with the behaviors and properties of *systems*. A system, broadly defined, is a set of entities that, through their interactions, relationships, or dependencies, form a unified whole. It is always defined in terms of its *boundary*, which determines the entities that are or are not part of the system. Entities lying outside the system then become part of the system's *environment*.

A system can exhibit *properties* that produce *behaviors* which are distinct from the properties and behaviors of its parts; these system-wide or *global* properties and behaviors are characteristics of how the system interacts with or appears to its environment, or of how its parts behave (say, in response to external stimuli) by virtue of being within the system. The notion of *behavior* implies that the study of systems is also concerned with processes that take place over time (or, in mathematics, some other phase space parameterization). Because of their broad, interdisciplinary applicability, systems concepts play a central role in complex systems.

As a field of study, complex systems is a subset of systems theory. General systems theory focuses similarly on the collective behaviors of interacting entities, but it studies a much broader class of systems, including non-complex systems where traditional reductionist approaches may remain viable. Indeed, systems theory seeks to explore and describe *all* classes of systems, and the invention of categories that are useful to researchers across widely varying fields is one of systems theory's main objectives.

As it relates to complex systems, systems theory contributes an emphasis on the way relationships and dependencies between a system's parts can determine system-wide properties. It also contributes the interdisciplinary perspective of the study of complex systems: the notion that shared properties link systems across disciplines, justifying the pursuit of modeling approaches applicable to complex systems wherever they appear. Specific concepts important to complex systems, such as emergence, feedback loops, and adaptation, also originate in systems theory.

Complexity

Systems exhibit complexity means that their behaviors cannot be easily implied from the very properties that make them difficult to model, and the complex behaviors are governed entirely, or almost entirely, by the behaviors those properties produce. Any modeling approach that ignores such difficulties or characterizes them as noise, then, will necessarily produce models that are neither accurate nor useful. As yet no fully general theory of complex systems has emerged for addressing these problems, so researchers must solve them in domain-specific contexts. Researchers in complex systems address these problems by viewing the chief task of modeling to be capturing, rather than reducing, the complexity of their respective systems of interest.

While no generally accepted exact definition of complexity exists yet, there are many archetypal examples of complexity. Systems can be complex if, for instance, they have chaotic behavior (behavior that exhibits extreme sensitivity to initial conditions), or if they have emergent properties (properties that are not apparent from their components in isolation but which result from the relationships and dependencies they form when placed together in a system), or if they are computationally intractable to model (if they depend on a number of parameters that grows too rapidly with respect to the size of the system).

Networks

The interacting components of a complex system form a network, which is a collection of discrete objects and relationships between them, usually depicted as a graph of vertices connected by edges. Networks can describe the relationships between individuals within an organization, between logic gates in a circuit, between genes in gene regulatory networks, or between any other set of related entities.

Networks often describe the sources of complexity in complex systems. Studying complex systems as networks therefore enables many useful applications of graph theory and network science. Some complex systems, for example, are also complex networks, which have properties such as phase transitions and power-law degree distributions that readily lend themselves to emergent or chaotic behavior. The fact that the number of edges in a complete graph grows quadratically in the number of vertices sheds additional light on the source of complexity in large networks: as a network grows, the number of relationships between entities quickly dwarfs the number of entities in the network.

Nonlinearity

Complex systems often have nonlinear behavior, meaning they may respond in different ways to the same input depending on their state or context. In mathematics and physics, nonlinearity describes systems in which a change in the size of the input does not produce a proportional change in the size of the output. For a given change in input, such systems may yield significantly greater than or less than proportional changes in output, or even no output at all, depending on the current state of the system or its parameter values.

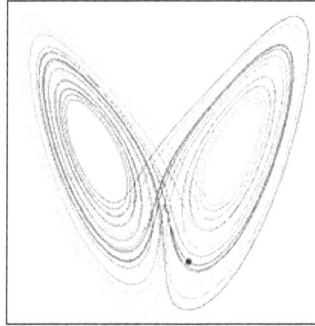

A sample solution in the Lorenz attractor when $\rho = 28$, $\sigma = 10$, and $\beta = 8/3$.

Of particular interest to complex systems are nonlinear dynamical systems, which are systems of differential equations that have one or more nonlinear terms. Some nonlinear dynamical systems, such as the Lorenz system, can produce a mathematical phenomenon known as chaos. Chaos as it applies to complex systems refers to the sensitive dependence on initial conditions, or "butterfly effect," that a complex system can exhibit. In such a system, small changes to initial conditions can lead to dramatically different outcomes. Chaotic behavior can therefore be extremely hard to model numerically, because small rounding errors at an intermediate stage of computation can cause the model to generate completely inaccurate output. Furthermore, if a complex system returns to a state similar to one it held previously, it may behave completely differently in response to exactly the same stimuli, so chaos also poses challenges for extrapolating from past experience.

Emergence

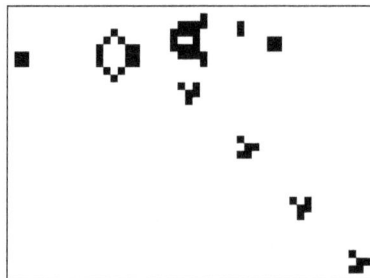

Gosper's Glider Gun creating "gliders" in the cellular automaton Conway's Game of Life.

Another common feature of complex systems is the presence of emergent behaviors and properties: these are traits of a system which are not apparent from its components in isolation but which result from the interactions, dependencies, or relationships they form when placed together in a system. Emergence broadly describes the appearance of such behaviors and properties, and has applications to systems studied in both the social and physical sciences. While emergence is often used to refer only to the appearance of unplanned organized behavior in a complex system, emergence can also refer to the breakdown of organization; it describes any phenomena which are difficult or even impossible to predict from the smaller entities that make up the system.

One example of complex system whose emergent properties have been studied extensively is cellular automata. In a cellular automaton, a grid of cells, each having one of finitely many states, evolves over time according to a simple set of rules. These rules guide the "interactions" of each cell with its neighbors. Although the rules are only defined locally, they have been shown capable of producing globally interesting behavior, for example in Conway's Game of Life.

Spontaneous Order and Self-organization

When emergence describes the appearance of unplanned order, it is spontaneous order (in the social sciences) or self-organization (in physical sciences). Spontaneous order can be seen in herd behavior, whereby a group of individuals coordinates their actions without centralized planning. Self-organization can be seen in the global symmetry of certain crystals, for instance the apparent radial symmetry of snowflakes, which arises from purely local attractive and repulsive forces both between water molecules and between water molecules and their surrounding environment.

Adaptation

Complex adaptive systems are special cases of complex systems that are adaptive in that they have the capacity to change and learn from experience. Examples of complex adaptive systems include the stock market, social insect and ant colonies, the biosphere and the ecosystem, the brain and the immune system, the cell and the developing embryo, the cities, manufacturing businesses and any human social group-based endeavor in a cultural and social system such as political parties or communities.

Features

Complex systems may have the following features:

Cascading Failures

Due to the strong coupling between components in complex systems, a failure in one or more components can lead to cascading failures which may have catastrophic consequences on the functioning of the system. Localized attack may lead to cascading failures and abrupt collapse in spatial networks.

Complex Systems may be Open

Complex systems are usually open systems — that is, they exist in a thermodynamic gradient and dissipate energy. In other words, complex systems are frequently far from energetic equilibrium: but despite this flux, there may be pattern stability, see synergetics.

Complex Systems may have a Memory

The history of a complex system may be important. Because complex systems are dynamical systems they change over time, and prior states may have an influence on present states. More formally, complex systems often exhibit spontaneous failures and recovery as well as hysteresis. Interacting systems may have complex hysteresis of many transitions.

Complex Systems may be Nested

The components of a complex system may themselves be complex systems. For example, an economy is made up of organisations, which are made up of people, which are made up of cells - all of which are complex systems.

Dynamic Network of Multiplicity

As well as coupling rules, the dynamic network of a complex system is important. Small-world or scale-free networks which have many local interactions and a smaller number of inter-area connections are often employed. Natural complex systems often exhibit such topologies. In the human cortex for example, we see dense local connectivity and a few very long axon projections between regions inside the cortex and to other brain regions.

May Produce Emergent Phenomena

Complex systems may exhibit behaviors that are emergent, which is to say that while the results may be sufficiently determined by the activity of the systems' basic constituents, they may have properties that can only be studied at a higher level. For example, the termites in a mound have physiology, biochemistry and biological development that are at one level of analysis, but their social behavior and mound building is a property that emerges from the collection of termites and needs to be analysed at a different level.

Relationships are Non-linear

In practical terms, this means a small perturbation may cause a large effect, a proportional effect, or even no effect at all. In linear systems, effect is always directly proportional to cause.

Relationships Contain Feedback Loops

Both negative (damping) and positive (amplifying) feedback are always found in complex systems. The effects of an element's behaviour are fed back to in such a way that the element itself is altered.

Applications

Complexity in Practice

The traditional approach to dealing with complexity is to reduce or constrain it. Typically, this involves compartmentalisation: dividing a large system into separate parts. Organizations, for instance, divide their work into departments that each deal with separate issues. Engineering systems are often designed using modular components. However, modular designs become susceptible to failure when issues arise that bridge the divisions.

Complexity Management

As projects and acquisitions become increasingly complex, companies and governments are challenged to find effective ways to manage mega-acquisitions such as the Army Future Combat Systems. Acquisitions such as the FCS rely on a web of interrelated parts which interact unpredictably. As acquisitions become more network-centric and complex, businesses will be forced to find ways to manage complexity while governments will be challenged to provide effective governance to ensure flexibility and resiliency.

Complexity Economics

Over the last decades, within the emerging field of complexity economics new predictive tools have been developed to explain economic growth. Such is the case with the models built by the Santa Fe Institute in 1989 and the more recent economic complexity index (ECI), introduced by the MIT physicist Cesar A. Hidalgo and the Harvard economist Ricardo Hausmann. Based on the ECI, Hausmann, Hidalgo and their team of The Observatory of Economic Complexity have produced GDP forecasts for the year 2020.

Complexity and Education

Focusing on issues of student persistence with their studies, Forsman, Moll and Linder explore the "viability of using complexity science as a frame to extend methodological applications for physics education research", finding that "framing a social network analysis within a complexity science perspective offers a new and powerful applicability across a broad range of PER topics".

Complexity and Modeling

One of Friedrich Hayek's main contributions to early complexity theory is his distinction between the human capacity to predict the behaviour of simple systems and its capacity to predict the behaviour of complex systems through modeling. He believed that economics and the sciences of complex phenomena in general, which in his view included biology, psychology, and so on, could not be modeled after the sciences that deal with essentially simple phenomena like physics. Hayek would notably explain that complex phenomena, through modeling, can only allow pattern predictions, compared with the precise predictions that can be made out of non-complex phenomena.

Complexity and Chaos Theory

Complexity theory is rooted in chaos theory, which in turn has its origins more than a century ago in the work of the French mathematician Henri Poincaré. Chaos is sometimes viewed as extremely complicated information, rather than as an absence of order. Chaotic systems remain deterministic, though their long-term behavior can be difficult to predict with any accuracy. With perfect knowledge of the initial conditions and of the relevant equations describing the chaotic system's behavior, one can theoretically make perfectly accurate predictions about the future of the system, though in practice this is impossible to do with arbitrary accuracy. Ilya Prigogine argued that complexity is non-deterministic, and gives no way whatsoever to precisely predict the future.

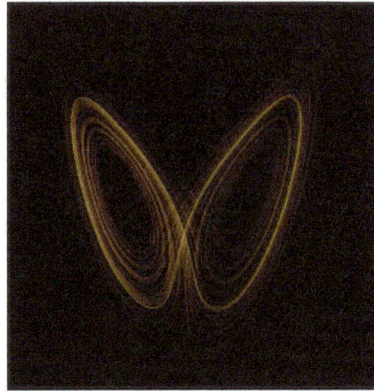

A plot of the Lorenz attractor.

The emergence of complexity theory shows a domain between deterministic order and randomness which is complex. This is referred as the "edge of chaos".

When one analyzes complex systems, sensitivity to initial conditions, for example, is not an issue as important as it is within chaos theory, in which it prevails. As stated by Colander, the study of complexity is the opposite of the study of chaos. Complexity is about how a huge number of extremely complicated and dynamic sets of relationships can generate some simple behavioral patterns, whereas chaotic behavior, in the sense of deterministic chaos, is the result of a relatively small number of non-linear interactions.

Therefore, the main difference between chaotic systems and complex systems is their history. Chaotic systems do not rely on their history as complex ones do. Chaotic behaviour pushes a system in equilibrium into chaotic order, which means, in other words, out of what we traditionally define as 'order'. On the other hand, complex systems evolve far from equilibrium at the edge of chaos. They evolve at a critical state built up by a history of irreversible and unexpected events, which physicist Murray Gell-Mann called "an accumulation of frozen accidents". In a sense chaotic systems can be regarded as a subset of complex systems distinguished precisely by this absence of historical dependence. Many real complex systems are, in practice and over long but finite time periods, robust. However, they do possess the potential for radical qualitative change of kind whilst retaining systemic integrity. Metamorphosis serves as perhaps more than a metaphor for such transformations.

Complexity and Network Science

A complex system is usually composed of many components and their interactions. Such a system can be represented by a network where nodes represent the components and links represent their interactions. for example, the INTERNET can be represented as a network composed of nodes (computers) and links (direct connections between computers). Its resilience to failures was studied using percolation theory. Other examples are social networks, airline networks, biological networks and climate networks. Networks can also fail and recover spontaneously. Traffic in a city can be represented as a network. The weighted links represent the velocity between two junctions (nodes). This approach was found useful to characterize the global traffic efficiency in a city. The complex pattern of exposures between financial institutions has been shown to trigger financial instability.

General form of Complexity Computation

The computational law of reachable optimality is established as a general form of computation for ordered systems and it reveals complexity computation is a compound computation of optimal choice and optimality driven reaching pattern over time underlying a specific and any experience path of ordered system within the general limitation of system integrity.

The computational law of reachable optimality has four key components as described below:

- Reachability of Optimality: Any intended optimality shall be reachable. Unreachable optimality has no meaning for a member in the ordered system and even for the ordered system itself.

- Prevailing and Consistency: Maximizing reachability to explore best available optimality is the prevailing computation logic for all members in the ordered system and is accommodated by the ordered system.

- Conditionality: Realizable tradeoff between reachability and optimality depends primarily upon the initial bet capacity and how the bet capacity evolves along with the payoff table update path triggered by bet behavior and empowered by the underlying law of reward and punishment. Precisely, it is a sequence of conditional events where the next event happens upon reached status quo from experience path.

- Robustness: The more challenge a reachable optimality can accommodate, the more robust it is in term of path integrity.

There are also four computation features in the law of reachable optimality:

- Optimal Choice: Computation in realizing Optimal Choice can be very simple or very complex. A simple rule in Optimal Choice is to accept whatever is reached, Reward As You Go (RAYG). A Reachable Optimality computation reduces into optimizing reachability when RAYG is adopted. The Optimal Choice computation can be more complex when multiple NE strategies present in a reached game.

- Initial Status: Computation is assumed to start at an interested beginning even the absolute beginning of an ordered system in nature may not and need not present. An assumed neutral Initial Status facilitates an artificial or a simulating computation and is not expected to change the prevalence of any findings.

- Territory: An ordered system shall have a territory where the universal computation sponsored by the system will produce an optimal solution still within the territory.

- Reaching Pattern: The forms of Reaching Pattern in the computation space, or the Optimality Driven Reaching Pattern in the computation space, primarily depend upon the nature and dimensions of measure space underlying a computation space and the law of punishment and reward underlying the realized experience path of reaching. There are five basic forms of experience path we are interested in, persistently positive reinforcement experience path, persistently negative reinforcement experience path, mixed persistent pattern experience path, decaying scale experience path and selection experience path.

The compound computation in selection experience path includes current and lagging interaction, dynamic topological transformation and implies both invariance and variance characteristics in an ordered system's experience path.

In addition, the computation law of reachable optimality gives out the boundary between complexity model, chaotic model and determination model. When RAYG is the Optimal Choice computation, and the reaching pattern is a persistently positive experience path, persistently negative experience path, or mixed persistent pattern experience path, the underlying computation shall be a simple system computation adopting determination rules. If the reaching pattern has no persistent pattern experienced in RAYG regime, the underlying computation hints there is a chaotic system. When the optimal choice computation involves non-RAYG computation, it's a complexity computation driving the compound effect.

COMPLEX ADAPTIVE SYSTEM

A complex adaptive system is a system in which a perfect understanding of the individual parts does not automatically convey a perfect understanding of the whole system's behavior. The study of complex adaptive systems, a subset of nonlinear dynamical systems, is highly interdisciplinary and blends insights from the natural and social sciences to develop system-level models and insights that allow for heterogeneous agents, phase transition, and emergent behavior.

They are *complex* in that they are dynamic networks of interactions, and their relationships are not aggregations of the individual static entities, i.e., the behavior of the ensemble is not predicted by the behavior of the components. They are *adaptive* in that the individual and collective behavior mutate and self-organize corresponding to the change-initiating micro-event or collection of events. They are a "complex macroscopic collection" of relatively "similar and partially connected micro-structures" formed in order to adapt to the changing environment and increase their survivability as a macro-structure.

The term *complex adaptive systems*, or *complexity science*, is often used to describe the loosely organized academic field that has grown up around the study of such systems. Complexity science is not a single theory—it encompasses more than one theoretical framework and is highly interdisciplinary, seeking the answers to some fundamental questions about living, adaptable, changeable systems. Complex adaptive systems may adopt hard or softer approaches. Hard theories use formal language that is precise, tend to see agents as having tangible properties, and usually see objects in a behavioral system that can be manipulated in some way. Softer theories use natural language and narratives that may be imprecise, and agents are subjects having both tangible and intangible properties. Examples of hard complexity theories include Complex Adaptive Systems (CAS) and Viability Theory, and a class of softer theory is Viable System Theory. Many of the propositional consideration made in hard theory are also of relevance to softer theory. From here on, interest will now center on CAS.

The study of CAS focuses on complex, emergent and macroscopic properties of the system. John H. Holland said that CAS "are systems that have a large numbers of components, often called agents, that interact and adapt or learn."

Typical examples of complex adaptive systems include: climate; cities; firms; markets; governments; industries; ecosystems; social networks; power grids; animal swarms; traffic flows; social insect (e.g. ant) colonies; the brain and the immune system; and the cell and the developing embryo. Human social group-based endeavors, such as political parties, communities, geopolitical organizations, war, and terrorist networks are also considered CAS. The internet and cyberspace—composed, collaborated, and managed by a complex mix of human–computer interactions, is also regarded as a complex adaptive system. CAS can be hierarchical, but more often exhibit aspects of "self-organization."

General Properties

What distinguishes a CAS from a pure multi-agent system (MAS) is the focus on top-level properties and features like self-similarity, complexity, emergence and self-organization. A MAS is defined as a system composed of multiple interacting agents; whereas in CAS, the agents as well as the system are adaptive and the system is self-similar. A CAS is a complex, self-similar collectivity of interacting, adaptive agents. Complex Adaptive Systems are characterized by a high degree of adaptive capacity, giving them resilience in the face of perturbation.

Other important properties are adaptation (or homeostasis), communication, cooperation, specialization, spatial and temporal organization, and reproduction. They can be found on all levels: cells specialize, adapt and reproduce themselves just like larger organisms do. Communication and cooperation take place on all levels, from the agent to the system level. The forces driving co-operation between agents in such a system, in some cases, can be analyzed with game theory.

Characteristics

Some of the most important characteristics of complex systems are:

- The number of elements is sufficiently large that conventional descriptions (e.g. a system of differential equations) are not only impractical, but cease to assist in understanding the system. Moreover, the elements interact dynamically, and the interactions can be physical or involve the exchange of information.

- Such interactions are rich, i.e. any element or sub-system in the system is affected by and affects several other elements or sub-systems.

- The interactions are non-linear: small changes in inputs, physical interactions or stimuli can cause large effects or very significant changes in outputs.

- Interactions are primarily but not exclusively with immediate neighbours and the nature of the influence is modulated.

- Any interaction can feed back onto itself directly or after a number of intervening stages. Such feedback can vary in quality. This is known as *recurrency*.

- The overall behavior of the system of elements is not predicted by the behavior of the individual elements.

- Such systems may be open and it may be difficult or impossible to define system boundaries.

- Complex systems operate under far from equilibrium conditions. There has to be a constant flow of energy to maintain the organization of the system.

- Complex systems have a history. They evolve and their past is co-responsible for their present behaviour.

- Elements in the system may be ignorant of the behaviour of the system as a whole, responding only to the information or physical stimuli available to them locally.

Robert Axelrod & Michael D. Cohen identify a series of key terms from a modeling perspective:

- Strategy, a conditional action pattern that indicates what to do in which circumstances.

- Artifact, a material resource that has definite location and can respond to the action of agents.

- Agent, a collection of properties, strategies & capabilities for interacting with artifacts & other agents.

- Population, a collection of agents, or, in some situations, collections of strategies.

- System, a larger collection, including one or more populations of agents and possibly also artifacts.

- Type, all the agents (or strategies) in a population that have some characteristic in common.

- Variety, the diversity of types within a population or system.

- Interaction pattern, the recurring regularities of contact among types within a system.

- Space (physical), location in geographical space & time of agents and artifacts.

- Space (conceptual), "location" in a set of categories structured so that "nearby" agents will tend to interact.

- Selection, processes that lead to an increase or decrease in the frequency of various types of agent or strategies.

- Success criteria or performance measures, a "score" used by an agent or designer in attributing credit in the selection of relatively successful (or unsuccessful) strategies or agents.

Modeling and Simulation

CAS are occasionally modeled by means of agent-based models and complex network-based models. Agent-based models are developed by means of various methods and tools primarily by means of first identifying the different agents inside the model. Another method of developing models for CAS involves developing complex network models by means of using interaction data of various CAS components.

CRITICAL SYSTEM

A critical system is a system which must be highly reliable and retain this reliability as they evolve without incurring prohibitive costs.

There are four types of critical systems: safety critical, mission critical, business critical and security critical.

For such systems, trusted methods and techniques must be used for development. Consequently, critical systems are usually developed using well-tested techniques rather than newer techniques that have not been subject to extensive practical experience. Developers of critical systems are naturally conservative, preferring to use older techniques whose strengths and weaknesses are understood, rather than new techniques which may appear to be better, but whose long-term problems are unknown.

Expensive software engineering techniques that are not cost-effective for non-critical systems may sometimes be used for critical systems development. For example, formal mathematical methods of software development have been successfully used for safety and security critical systems. One reason why these formal methods are used is that it helps reduce the amount of testing required. For critical systems, the costs of verification and validation are usually very high—more than 50% of the total system development costs.

Classification

A critical system is distinguished by the consequences associated with system or function failure. Likewise, critical systems are further distinguished between fail-operational and fail safe systems, according to the tolerance they must exhibit to failures:

- Fail-operational — typically required to operate not only in nominal conditions (expected), but also in degraded situations when some parts are not working properly. For example, airplanes are fail-operational because they must be able to fly even if some components fail.

- Fail-safe — must safely shut down in case of single or multiple failures. Trains are fail-safe systems because stopping a train is typically sufficient to put into safe state.

Safety Critical

Safety critical systems deal with scenarios that may lead to loss of life, serious personal injury, or damage to the natural environment. Examples of safety-critical systems are a control system for a chemical manufacturing plant, aircraft, the controller of an unmanned train metro system, a controller of a nuclear plant, etc.

Mission critical

Mission critical systems are made to avoid inability to complete the overall system, project objectives or one of the goals for which the system was designed. Examples of mission-critical systems are a navigational system for a spacecraft, software controlling a baggage handling system of an airport, etc.

Business Critical

Business critical systems are programmed to avoid significant tangible or intangible economic costs; e.g., loss of business or damage to reputation. This is often due to the interruption of service caused by the system being unusable. Examples of a business-critical systems are the customer accounting system in a bank, stock-trading system, ERP system of a company, Internet search engine, etc.

Security Critical

Security critical systems deal with the loss of sensitive data through theft or accidental loss.

References

- Electronic-system, systems: electronics-tutorials.ws, Retrieved 8 January, 2019

- 1936-, radin, beryl a. (2000). Beyond machiavelli: policy analysis comes of age. Washington, d.c.: georgetown university press. Isbn 0878407731. Oclc 41834855

- 05-systemmodeling, notes16, cs410, classes, stan: ccsu.edu, Retrieved 9 January, 2019

- Bozzano, marco; villafiorita, adolfo (2010). Design and safety assessment of critical systemss. Austin, texas: auerbach publications. P. 298. Isbn 9781439803318

Control systems are used to manage, direct, command or regulate the behavior of other systems or devices using control loops. Some of the different types of control systems are open loop control systems and feedback control systems. This chapter discusses in detail these concepts related to control systems as well as the applications of control theory in biomedical engineering.

A control system is a system, which provides the desired response by controlling the output. The following figure shows the simple block diagram of a control system.

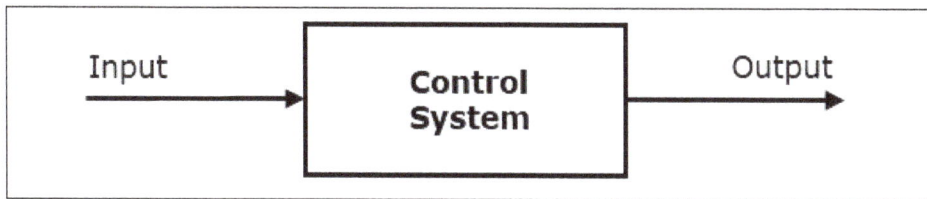

Here, the control system is represented by a single block. Since, the output is controlled by varying input, the control system got this name. We will vary this input with some mechanism.

Examples – Traffic lights control system, washing machine.

Traffic lights control system is an example of control system. Here, a sequence of input signal is applied to this control system and the output is one of the three lights that will be on for some duration of time. During this time, the other two lights will be off. Based on the traffic study at a particular junction, the on and off times of the lights can be determined. Accordingly, the input signal controls the output. So, the traffic lights control system operates on time basis.

Classification of Control Systems

Based on some parameters, we can classify the control systems into the following ways:

Continuous Time and Discrete-time Control Systems

- Control Systems can be classified as continuous time control systems and discrete time control systems based on the type of the signal used.

- In continuous time control systems, all the signals are continuous in time. But, in discrete time control systems, there exists one or more discrete time signals.

SISO and MIMO Control Systems

- Control Systems can be classified as SISO control systems and MIMO control systems based on the number of inputs and outputs present.

- SISO (Single Input and Single Output) control systems have one input and one output. Whereas, MIMO (Multiple Inputs and Multiple Outputs) control systems have more than one input and more than one output.

Open Loop and Closed Loop Control Systems

Control Systems can be classified as open loop control systems and closed loop control systems based on the feedback path.

In open loop control systems, output is not fed-back to the input. So, the control action is independent of the desired output.

The following figure shows the block diagram of the open loop control system:

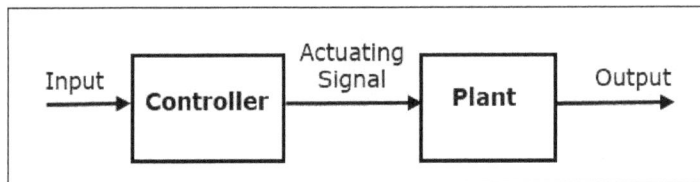

Here, an input is applied to a controller and it produces an actuating signal or controlling signal. This signal is given as an input to a plant or process which is to be controlled. So, the plant produces an output, which is controlled. The traffic lights control system is an example of an open loop control system.

In closed loop control systems, output is fed back to the input. So, the control action is dependent on the desired output.

The following figure shows the block diagram of negative feedback closed loop control system:

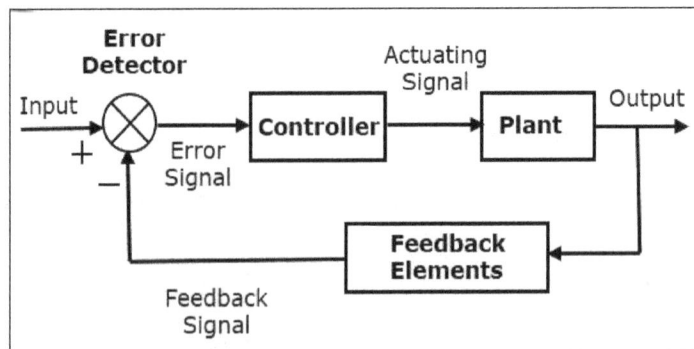

The error detector produces an error signal, which is the difference between the input and the feedback signal. This feedback signal is obtained from the block (feedback elements) by considering the output of the overall system as an input to this block. Instead of the direct input, the error signal is applied as an input to a controller.

So, the controller produces an actuating signal which controls the plant. In this combination, the output of the control system is adjusted automatically till we get the desired response. Hence, the closed loop control systems are also called the automatic control systems. Traffic lights control system having sensor at the input is an example of a closed loop control system.

The differences between the open loop and the closed loop control systems are mentioned in the following table:

Open Loop Control Systems	Closed Loop Control Systems
Control action is independent of the desired output.	Control action is dependent of the desired output.
Feedback path is not present.	Feedback path is present.
These are also called as non-feedback control systems.	These are also called as feedback control systems.
Easy to design.	Difficult to design.
These are economical.	These are costlier.
Inaccurate.	Accurate.

CONTROL THEORY

Control theory in control systems engineering is a subfield of mathematics that deals with the control of continuously operating dynamical systems in engineered processes and machines. The objective is to develop a control model for controlling such systems using a control action in an optimum manner without *delay or overshoot* and ensuring control stability.

To do this, a *controller* with the requisite corrective behaviour is required. This controller monitors the controlled process variable (PV), and compares it with the reference or set point (SP). The difference between actual and desired value of the process variable, called the *error* signal, or SP-PV error, is applied as feedback to generate a control action to bring the controlled process variable to the same value as the set point. Other aspects which are also studied are controllability and observability. On this is based the advanced type of automation that revolutionized manufacturing, aircraft, communications and other industries. This is *feedback control*, which is usually *continuous* and involves taking measurements using a sensor and making calculated adjustments to keep the measured variable within a set range by means of a "final control element", such as a control valve.

Extensive use is usually made of a diagrammatic style known as the block diagram. In it the transfer function, also known as the system function or network function, is a mathematical model of the relation between the input and output based on the differential equations describing the system.

Control theory dates from the 19th century, when the theoretical basis for the operation of governors was first described by James Clerk Maxwell. Control theory was further advanced by Edward Routh in 1874, Charles Sturm and in 1895, Adolf Hurwitz, who all contributed to the establishment of control stability criteria; and from 1922 onwards, the development of PID control theory by Nicolas Minorsky. Although a major application of control theory is in control systems engineering, which deals with the design of process control systems for industry, other applications range far beyond this. As the general theory of feedback systems, control theory is useful wherever feedback occurs.

Although control systems of various types date back to antiquity, a more formal analysis of the field began with a dynamics analysis of the centrifugal governor, conducted by the physicist James Clerk Maxwell in 1868, entitled *On Governors*. A centrifugal governor was already used to regulate the velocity of windmills. Maxwell described and analyzed the phenomenon of self-oscillation, in which lags in the system may lead to overcompensation and unstable behavior. This generated a flurry of interest in the topic, during which Maxwell's classmate, Edward John Routh, abstracted Maxwell's results for the general class of linear systems. Independently, Adolf Hurwitz analyzed system stability using differential equations in 1877, resulting in what is now known as the Routh–Hurwitz theorem.

A notable application of dynamic control was in the area of manned flight. The Wright brothers made their first successful test flights on December 17, 1903 and were distinguished by their ability to control their flights for substantial periods (more so than the ability to produce lift from an airfoil, which was known). Continuous, reliable control of the airplane was necessary for flights lasting longer than a few seconds.

By World War II, control theory was becoming an important area of research. Irmgard Flügge-Lotz developed the theory of discontinuous automatic control systems, and applied the bang-bang principle to the development of automatic flight control equipment for aircraft. Other areas of application for discontinuous controls included fire-control systems, guidance systems and electronics.

Sometimes, mechanical methods are used to improve the stability of systems. For example, ship stabilizers are fins mounted beneath the waterline and emerging laterally. In contemporary vessels, they may be gyroscopically controlled active fins, which have the capacity to change their angle of attack to counteract roll caused by wind or waves acting on the ship.

The Space Race also depended on accurate spacecraft control, and control theory has also seen an increasing use in fields such as economics.

Classical Control Theory

To overcome the limitations of the open-loop controller, control theory introduces feedback. A closed-loop controller uses feedback to control states or outputs of a dynamical system. Its name comes from the information path in the system: process inputs (e.g., voltage applied to an electric motor) have an effect on the process outputs (e.g., speed or torque of the motor), which is measured with sensors and processed by the controller; the result (the control signal) is "fed back" as input to the process, closing the loop.

Closed-loop controllers have the following advantages over open-loop controllers:

- Disturbance rejection (such as hills in the cruise control).

- Guaranteed performance even with model uncertainties, when the model structure does not match perfectly the real process and the model parameters are not exact.

- Unstable processes can be stabilized.

- Reduced sensitivity to parameter variations.

- Improved reference tracking performance.

In some systems, closed-loop and open-loop control are used simultaneously. In such systems, the open-loop control is termed feedforward and serves to further improve reference tracking performance. A common closed-loop controller architecture is the PID controller.

Closed-loop Transfer Function

The output of the system *y(t)* is fed back through a sensor measurement *F* to a comparison with the reference value *r(t)*. The controller *C* then takes the error *e* (difference) between the reference and the output to change the inputs *u* to the system under control *P*. This is shown in the figure. This kind of controller is a closed-loop controller or feedback controller.

This is called a single-input-single-output (*SISO*) control system; *MIMO* (i.e., Multi-Input-Multi-Output) systems, with more than one input/output, are common. In such cases variables are represented through vectors instead of simple scalar values. For some distributed parameter systems the vectors may be infinite-dimensional (typically functions).

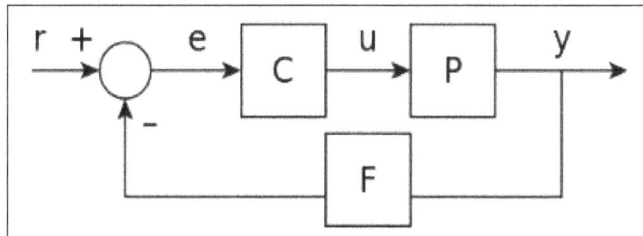

If we assume the controller *C*, the plant *P*, and the sensor *F* are linear and time-invariant (i.e., elements of their transfer function *C(s)*, *P(s)*, and *F(s)* do not depend on time), the systems above can be analysed using the Laplace transform on the variables. This gives the following relations:

$$Y(s) = P(s)U(s)$$

$$U(s) = C(s)E(s)$$

$$E(s) = R(s) - F(s)Y(s).$$

Solving for *Y(s)* in terms of *R(s)* gives,

$$Y(s) = \left(\frac{P(s)C(s)}{1 + P(s)C(s)F(s)} \right) R(s) = H(s)R(s).$$

The expression $H(s) = \dfrac{P(s)C(s)}{1 + F(s)P(s)C(s)}$ is referred to as the *closed-loop transfer function* of the

system. The numerator is the forward (open-loop) gain from *r* to *y*, and the denominator is one plus the gain in going around the feedback loop, the so-called loop gain. If $|P(s)C(s)| \gg 1$, i.e., it has a large norm with each value of *s*, and if $|F(s)| \approx 1$, then *Y(s)* is approximately equal to *R(s)* and the output closely tracks the reference input.

PID Feedback Control

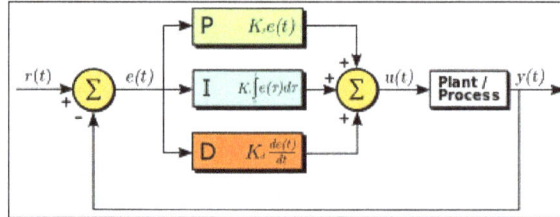

A block diagram of a PID controller in a feedback loop, r(*t*) is the desired process value or "set point", and y(*t*) is the measured process value.

A proportional–integral–derivative controller (PID controller) is a control loop feedback mechanism control technique widely used in control systems.

A PID controller continuously calculates an *error value* $e(t)$ as the difference between a desired setpoint and a measured process variable and applies a correction based on proportional, integral, and derivative terms. *PID* is an initialism for *Proportional-Integral-Derivative*, referring to the three terms operating on the error signal to produce a control signal.

The theoretical understanding and application dates from the 1920s, and they are implemented in nearly all analogue control systems; originally in mechanical controllers, and then using discrete electronics and latterly in industrial process computers. The PID controller is probably the most-used feedback control design.

If *u(t)* is the control signal sent to the system, *y(t)* is the measured output and *r(t)* is the desired output, and $e(t) = r(t) - y(t)$ is the tracking error, a PID controller has the general form,

$$u(t) = K_P e(t) + K_I \int e(\tau) \mathrm{d}\tau + K_D \frac{\mathrm{d}e(t)}{\mathrm{d}t}.$$

The desired closed loop dynamics is obtained by adjusting the three parameters K_P, K_I and K_D, often iteratively by "tuning" and without specific knowledge of a plant model. Stability can often be ensured using only the proportional term. The integral term permits the rejection of a step disturbance (often a striking specification in process control). The derivative term is used to provide damping or shaping of the response. PID controllers are the most well-established class of control systems: however, they cannot be used in several more complicated cases, especially if MIMO systems are considered.

Applying Laplace transformation results in the transformed PID controller equation,

$$u(s) = K_P e(s) + K_I \frac{1}{s} e(s) + K_D s e(s)$$

$$u(s) = \left(K_P + K_I \frac{1}{s} + K_D s \right) e(s)$$

with the PID controller transfer function,

$$C(s) = \left(K_P + K_I \frac{1}{s} + K_D s \right).$$

As an example of tuning a PID controller in the closed-loop system $H(s)$, consider a 1st order plant given by,

$$P(s) = \frac{A}{1 + sT_P}$$

where A and T_P are some constants. The plant output is fed back through,

$$F(s) = \frac{1}{1 + sT_F}$$

where T_F is also a constant. Now if we set $K_P = K\left(1 + \frac{T_D}{T_I}\right)$, $K_D = KT_D$, and $K_I = \frac{K}{T_I}$, we can express the PID controller transfer function in series form as,

$$C(s) = K\left(1 + \frac{1}{sT_I}\right)(1 + sT_D)$$

Plugging $P(s)$, $F(s)$, and $C(s)$ into the closed-loop transfer function $H(s)$, we find that by setting,

$$K = \frac{1}{A}, T_I = T_F, T_D = T_P$$

$H(s) = 1$. With this tuning in this example, the system output follows the reference input exactly.

However, in practice, a pure differentiator is neither physically realizable nor desirable due to amplification of noise and resonant modes in the system. Therefore, a phase-lead compensator type approach or a differentiator with low-pass roll-off are used instead.

Linear and Nonlinear Control Theory

The field of control theory can be divided into two branches:

- *Linear control theory* – This applies to systems made of devices which obey the superposition principle, which means roughly that the output is proportional to the input. They are governed by linear differential equations. A major subclass is systems which in addition have parameters which do not change with time, called *linear time invariant* (LTI) systems. These systems are amenable to powerful frequency domain mathematical techniques of great generality, such as the Laplace transform, Fourier transform, Z transform, Bode plot, root locus, and Nyquist stability criterion. These lead to a description of the system using terms like bandwidth, frequency response, eigenvalues, gain, resonant frequencies, zeros and poles, which give solutions for system response and design techniques for most systems of interest.

- *Nonlinear control theory* – This covers a wider class of systems that do not obey the superposition principle, and applies to more real-world systems because all real control systems are nonlinear. These systems are often governed by nonlinear differential equations. The few mathematical techniques which have been developed to handle them are more difficult and much less general, often applying only to narrow categories of systems. These include limit cycle theory, Poincaré maps, Lyapunov stability theorem, and describing functions. Nonlinear systems are often analyzed using numerical methods on computers, for example by simulating their operation using a simulation language. If only solutions near a stable point are of interest, nonlinear systems can often be linearized by approximating them by a linear system using perturbation theory, and linear techniques can be used.

Analysis Techniques - frequency Domain and Time Domain

Mathematical techniques for analyzing and designing control systems fall into two different categories:

- *Frequency domain* – In this type the values of the state variables, the mathematical variables representing the system's input, output and feedback are represented as functions of frequency. The input signal and the system's transfer function are converted from time functions to functions of frequency by a transform such as the Fourier transform, Laplace transform, or Z transform. The advantage of this technique is that it results in a simplification of the mathematics; the *differential equations* that represent the system are replaced by *algebraic equations* in the frequency domain which is much simpler to solve. However, frequency domain techniques can only be used with linear systems.

- *Time-domain state space representation* – In this type the values of the state variables are represented as functions of time. With this model, the system being analyzed is represented by one or more differential equations. Since frequency domain techniques are limited to linear systems, time domain is widely used to analyze real-world nonlinear systems. Although these are more difficult to solve, modern computer simulation techniques such as simulation languages have made their analysis routine.

In contrast to the frequency domain analysis of the classical control theory, modern control theory utilizes the time-domain state space representation, a mathematical model of a physical system as a set of input, output and state variables related by first-order differential equations. To abstract from the number of inputs, outputs, and states, the variables are expressed as vectors and the differential and algebraic equations are written in matrix form (the latter only being possible when the dynamical system is linear). The state space representation (also known as the "time-domain approach") provides a convenient and compact way to model and analyze systems with multiple inputs and outputs. With inputs and outputs, we would otherwise have to write down Laplace transforms to encode all the information about a system. Unlike the frequency domain approach, the use of the state-space representation is not limited to systems with linear components and zero initial conditions. "State space" refers to the space whose axes are the state variables. The state of the system can be represented as a point within that space.

System Interfacing - SISO and MIMO

Control systems can be divided into different categories depending on the number of inputs and outputs.

- Single-input single-output (SISO): This is the simplest and most common type, in which one output is controlled by one control signal. Examples are the cruise control, or an audio system, in which the control input is the input audio signal and the output is the sound waves from the speaker.

- Multiple-input multiple-output (MIMO): These are found in more complicated systems. For example, modern large telescopes such as the Keck and MMT have mirrors composed of many separate segments each controlled by an actuator. The shape of the entire mirror is constantly adjusted by a MIMO active optics control system using input from multiple sensors at the focal plane, to compensate for changes in the mirror shape due to thermal expansion, contraction, stresses as it is rotated and distortion of the wavefront due to turbulence in the atmosphere. Complicated systems such as nuclear reactors and human cells are simulated by a computer as large MIMO control systems.

Topics in Control Theory

Stability

The *stability* of a general dynamical system with no input can be described with Lyapunov stability criteria:

- A linear system is called bounded-input bounded-output (BIBO) stable if its output will stay bounded for any bounded input.

- Stability for nonlinear systems that take an input is input-to-state stability (ISS), which combines Lyapunov stability and a notion similar to BIBO stability.

For simplicity, the following descriptions focus on continuous-time and discrete-time linear systems.

Mathematically, this means that for a causal linear system to be stable all of the poles of its transfer function must have negative-real values, i.e. the real part of each pole must be less than zero. Practically speaking, stability requires that the transfer function complex poles reside

- In the open left half of the complex plane for continuous time, when the Laplace transform is used to obtain the transfer function.

- Inside the unit circle for discrete time, when the Z-transform is used.

The difference between the two cases is simply due to the traditional method of plotting continuous time versus discrete time transfer functions. The continuous Laplace transform is in Cartesian coordinates where the x axis is the real axis and the discrete Z-transform is in circular coordinates where the ρ axis is the real axis.

When the appropriate conditions above are satisfied a system is said to be asymptotically stable; the variables of an asymptotically stable control system always decrease from their initial value and do not show permanent oscillations. Permanent oscillations occur when a pole has a real part exactly equal to zero (in the continuous time case) or a modulus equal to one (in the discrete time case). If a simply stable system response neither decays nor grows over time, and has no oscillations, it is marginally stable; in this case the system transfer function has non-repeated poles at the complex plane origin (i.e. their real and complex component is zero in the continuous time case). Oscillations are present when poles with real part equal to zero have an imaginary part not equal to zero.

If a system in question has an impulse response of,

$$x[n] = 0.5^n u[n]$$

then the Z-transform is given by,

$$X(z) = \frac{1}{1 - 0.5z^{-1}}$$

which has a pole in $z = 0.5$ (zero imaginary part). This system is BIBO (asymptotically) stable since the pole is *inside* the unit circle.

However, if the impulse response was,

$$x[n] = 1.5^n u[n]$$

then the Z-transform is,

$$X(z) = \frac{1}{1 - 1.5z^{-1}}$$

which has a pole at $z = 1.5$ and is not BIBO stable since the pole has a modulus strictly greater than one.

Numerous tools exist for the analysis of the poles of a system. These include graphical systems like the root locus, Bode plots or the Nyquist plots.

Mechanical changes can make equipment (and control systems) more stable. Sailors add ballast to improve the stability of ships. Cruise ships use antiroll fins that extend transversely from the side of the ship for perhaps 30 feet (10 m) and are continuously rotated about their axes to develop forces that oppose the roll.

Controllability and Observability

Controllability and observability are main issues in the analysis of a system before deciding the best control strategy to be applied, or whether it is even possible to control or stabilize the system. Controllability is related to the possibility of forcing the system into a particular state by using an appropriate control signal. If a state is not controllable, then no signal will ever be able to control the state. If a state is not controllable, but its dynamics are stable, then the state is termed *stabilizable*. Observability instead is related to the possibility of *observing*, through output measurements, the state of a system. If a state is not observable, the controller will never be able to determine the behavior of an unobservable state and hence cannot use it to stabilize the system. However, similar to the stabilizability condition above, if a state cannot be observed it might still be detectable.

From a geometrical point of view, looking at the states of each variable of the system to be controlled, every "bad" state of these variables must be controllable and observable to ensure a good behavior in the closed-loop system. That is, if one of the eigenvalues of the system is not both controllable and observable, this part of the dynamics will remain untouched in the closed-loop system. If such an eigenvalue is not stable, the dynamics of this eigenvalue will be present in the closed-loop system which therefore will be unstable. Unobservable poles are not present in the transfer function realization of a state-space representation, which is why sometimes the latter is preferred in dynamical systems analysis.

Solutions to problems of an uncontrollable or unobservable system include adding actuators and sensors.

Control Specification

Several different control strategies have been devised in the past years. These vary from extremely general ones (PID controller), to others devoted to very particular classes of systems (especially robotics or aircraft cruise control).

A control problem can have several specifications. Stability, of course, is always present. The controller must ensure that the closed-loop system is stable, regardless of the open-loop stability. A poor choice of controller can even worsen the stability of the open-loop system, which must normally be avoided. Sometimes it would be desired to obtain particular dynamics in the closed loop: i.e. that the poles have $Re[\lambda] < -\bar{\lambda}$, where $\bar{\lambda}$ is a fixed value strictly greater than zero, instead of simply asking that $Re[\lambda] < 0$.

Another typical specification is the rejection of a step disturbance; including an integrator in the open-loop chain (i.e. directly before the system under control) easily achieves this. Other classes of disturbances need different types of sub-systems to be included.

Other "classical" control theory specifications regard the time-response of the closed-loop system. These include the rise time (the time needed by the control system to reach the desired value after a perturbation), peak overshoot (the highest value reached by the response before reaching the desired value) and others (settling time, quarter-decay). Frequency domain specifications are usually related to robustness.

Modern performance assessments use some variation of integrated tracking error (IAE,ISA,CQI).

Model Identification and Robustness

A control system must always have some robustness property. A robust controller is such that its properties do not change much if applied to a system slightly different from the mathematical one used for its synthesis. This requirement is important, as no real physical system truly behaves like the series of differential equations used to represent it mathematically. Typically a simpler mathematical model is chosen in order to simplify calculations, otherwise, the true system dynamics can be so complicated that a complete model is impossible.

System Identification

The process of determining the equations that govern the model's dynamics is called system identification. This can be done off-line: for example, executing a series of measures from which to calculate an approximated mathematical model, typically its transfer function or matrix. Such identification from the output, however, cannot take account of unobservable dynamics. Sometimes the model is built directly starting from known physical equations, for example, in the case of a mass-spring-damper system we know that $m\ddot{x}(t) = -Kx(t) - B\dot{x}(t)$. Even assuming that a "complete" model is used in designing the controller, all the parameters included in these equations (called "nominal parameters") are never known with absolute precision; the control system will have to behave correctly even when connected to a physical system with true parameter values away from nominal.

Some advanced control techniques include an "on-line" identification process. The parameters of the model are calculated ("identified") while the controller itself is running. In this way, if a drastic variation of the parameters ensues, for example, if the robot's arm releases a weight, the controller will adjust itself consequently in order to ensure the correct performance.

Analysis

Analysis of the robustness of a SISO (single input single output) control system can be performed in the frequency domain, considering the system's transfer function and using Nyquist and Bode diagrams. Topics include gain and phase margin and amplitude margin. For MIMO (multi-input multi output) and, in general, more complicated control systems, one must consider the theoretical results devised for each control technique. I.e., if particular robustness qualities are needed, the engineer must shift his attention to a control technique by including them in its properties.

Constraints

A particular robustness issue is the requirement for a control system to perform properly in the presence of input and state constraints. In the physical world every signal is limited. It could happen that a controller will send control signals that cannot be followed by the physical system, for example, trying to rotate a valve at excessive speed. This can produce undesired behavior of the closed-loop system, or even damage or break actuators or other subsystems. Specific control techniques are available to solve the problem: model predictive control, and anti-wind up systems. The latter consists of an additional control block that ensures that the control signal never exceeds a given threshold.

System Classifications

Linear Systems Control

For MIMO systems, pole placement can be performed mathematically using a state space representation of the open-loop system and calculating a feedback matrix assigning poles in the desired positions. In complicated systems this can require computer-assisted calculation capabilities, and cannot always ensure robustness. Furthermore, all system states are not in general measured and so observers must be included and incorporated in pole placement design.

Nonlinear Systems Control

Processes in industries like robotics and the aerospace industry typically have strong nonlinear dynamics. In control theory it is sometimes possible to linearize such classes of systems and apply linear techniques, but in many cases it can be necessary to devise from scratch theories permitting control of nonlinear systems. These, e.g., feedback linearization, backstepping, sliding mode control, trajectory linearization control normally take advantage of results based on Lyapunov's theory. Differential geometry has been widely used as a tool for generalizing well-known linear control concepts to the non-linear case, as well as showing the subtleties that make it a more challenging problem. Control theory has also been used to decipher the neural mechanism that directs cognitive states.

Decentralized Systems Control

When the system is controlled by multiple controllers, the problem is one of decentralized control. Decentralization is helpful in many ways, for instance, it helps control systems to operate over a larger geographical area. The agents in decentralized control systems can interact using communication channels and coordinate their actions.

Deterministic and Stochastic Systems Control

A stochastic control problem is one in which the evolution of the state variables is subjected to random shocks from outside the system. A deterministic control problem is not subject to external random shocks.

Main Control Strategies

Every control system must guarantee first the stability of the closed-loop behavior. For linear systems, this can be obtained by directly placing the poles. Non-linear control systems use specific theories (normally based on Aleksandr Lyapunov's Theory) to ensure stability without regard to the inner dynamics of the system. The possibility to fulfill different specifications varies from the model considered and the control strategy chosen.

List of the main control techniques:

- Adaptive control uses on-line identification of the process parameters, or modification of controller gains, thereby obtaining strong robustness properties. Adaptive controls were applied for the first time in the aerospace industry in the 1950s, and have found particular success in that field.

- A hierarchical control system is a type of control system in which a set of devices and governing software is arranged in a hierarchical tree. When the links in the tree are implemented by a computer network, then that hierarchical control system is also a form of networked control system.

- Intelligent control uses various AI computing approaches like artificial neural networks, Bayesian probability, fuzzy logic, machine learning, evolutionary computation and genetic algorithms to control a dynamic system.

- Optimal control is a particular control technique in which the control signal optimizes a certain "cost index": for example, in the case of a satellite, the jet thrusts needed to bring it to desired trajectory that consume the least amount of fuel. Two optimal control design methods have been widely used in industrial applications, as it has been shown they can guarantee closed-loop stability. These are Model Predictive Control (MPC) and linear-quadratic-Gaussian control (LQG). The first can more explicitly take into account constraints on the signals in the system, which is an important feature in many industrial processes. However, the "optimal control" structure in MPC is only a means to achieve such a result, as it does not optimize a true performance index of the closed-loop control system. Together with PID controllers, MPC systems are the most widely used control technique in process control.

- Robust control deals explicitly with uncertainty in its approach to controller design. Controllers designed using *robust control* methods tend to be able to cope with small differences between the true system and the nominal model used for design. The early methods of Bode and others were fairly robust; the state-space methods invented in the 1960s and 1970s were sometimes found to lack robustness. Examples of modern robust control techniques include H-infinity loop-shaping developed by Duncan McFarlane and Keith Glover, Sliding mode control (SMC) developed by Vadim Utkin, and safe protocols designed for control of large heterogeneous populations of electric loads in Smart Power Grid applications. Robust methods aim to achieve robust performance and/or stability in the presence of small modeling errors.

- Stochastic control deals with control design with uncertainty in the model. In typical stochastic control problems, it is assumed that there exist random noise and disturbances in the model and the controller, and the control design must take into account these random deviations.

- Energy-shaping control view the plant and the controller as energy-transformation devices. The control strategy is formulated in terms of interconnection (in a power-preserving manner) in order to achieve a desired behavior.

- Self-organized criticality control may be defined as attempts to interfere in the processes by which the self-organized system dissipates energy.

MATHEMATICAL MODELS OF SYSTEMS

The control systems can be represented with a set of mathematical equations known as mathematical model. These models are useful for analysis and design of control systems. Analysis of control system means finding the output when we know the input and mathematical model. Design of control system means finding the mathematical model when we know the input and the output.

The following mathematical models are mostly used:

- Differential equation model.
- Transfer function model.
- State space model.

Differential Equation Model

Differential equation model is a time domain mathematical model of control systems. Follow these steps for differential equation model:

- Apply basic laws to the given control system.
- Get the differential equation in terms of input and output by eliminating the intermediate variables.

Example:

Consider the following electrical system as shown in the following figure. This circuit consists of resistor, inductor and capacitor. All these electrical elements are connected in series. The input voltage applied to this circuit is v_i and the voltage across the capacitor is the output voltage v_o.

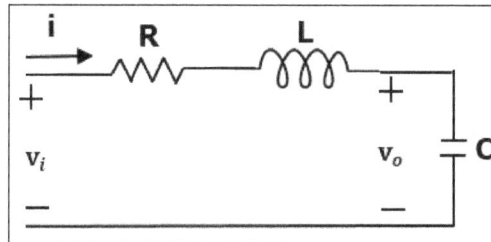

Mesh equation for this circuit is,

$$v_i = Ri + L\frac{di}{dt} + v_o$$

Substitute, the current passing through capacitor $i = c\frac{dv_o}{dt}$ in the above equation.

$$\Rightarrow v_i = RC\frac{dv_o}{dt} + LC\frac{d^2v_o}{dt^2} + v_o$$

$$\Rightarrow \frac{d^2v_o}{dt^2} + \left(\frac{R}{L}\right)\frac{dv_o}{dt} + \left(\frac{1}{LC}\right)v_o = \left(\frac{1}{LC}\right)v_i$$

The above equation is a second order differential equation.

Transfer Function Model

Transfer function model is an s-domain mathematical model of control systems. The Transfer function of a Linear Time Invariant (LTI) system is defined as the ratio of Laplace transform of output and Laplace transform of input by assuming all the initial conditions are zero.

If $x(t)$ and $y(t)$ are the input and output of an LTI system, then the corresponding Laplace transforms are $X(s)$ and $Y(s)$.

Therefore, the transfer function of LTI system is equal to the ratio of $Y(s)$ and $X(s)$.

$$i.e., Transfer\ function = \frac{Y(s)}{X(s)}$$

The transfer function model of an LTI system is shown in the following figure:

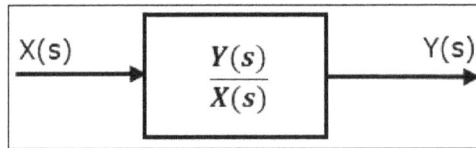

Here, we represented an LTI system with a block having transfer function inside it. And this block has an input $X(s)$ & output $Y(s)$.

Example:

Previously, we got the differential equation of an electrical system as,

$$\frac{d^2 v_o}{dt^2} + \left(\frac{R}{L}\right)\frac{dv_o}{dt} + \left(\frac{1}{LC}\right)v_o = \left(\frac{1}{LC}\right)v_i$$

Apply Laplace transform on both sides.

$$s^2 V_o(s) + \left(\frac{sR}{L}\right)V_o(s) + \left(\frac{1}{LC}\right)V_o(s) = \left(\frac{1}{LC}\right)v_i(s)$$

$$\Rightarrow \left\{ s^2 + \left(\frac{R}{L}\right)s + \frac{1}{LC} \right\} V_o(s) = \left(\frac{1}{LC}\right)V_i(s)$$

$$\Rightarrow \frac{V_o(s)}{V_i(s)} = \frac{\dfrac{1}{LC}}{s^2 + \left(\dfrac{R}{L}\right)s + \dfrac{1}{LC}}$$

where,

- $v_i(s)$ is the Laplace transform of the input voltage v_i.

- $v_o(s)$ is the Laplace transform of the output voltage v_o.

The above equation is a transfer function of the second order electrical system. The transfer function model of this system is shown.

$$\frac{1/LC}{s^2 + \left(\frac{R}{L}\right)s + \frac{1}{LC}}$$

$V_i(s)$ $V_o(s)$

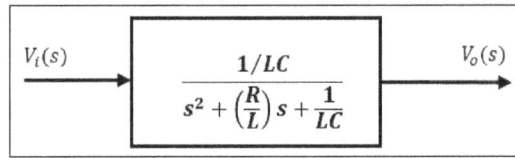

Here, we show a second order electrical system with a block having the transfer function inside it. And this block has an input $V_i(s)$ & an output $V_o(\text{s})$.

FEEDBACK SYSTEMS

Feedback Systems process signals and as such are signal processors. The processing part of a feedback system may be electrical or electronic, ranging from a very simple to a highly complex circuits.

Simple analogue feedback control circuits can be constructed using individual or discrete components, such as transistors, resistors and capacitors, etc, or by using microprocessor-based and integrated circuits (IC's) to form more complex digital feedback systems.

As we have seen, open-loop systems are just that, open ended, and no attempt is made to compensate for changes in circuit conditions or changes in load conditions due to variations in circuit parameters, such as gain and stability, temperature, supply voltage variations and/or external disturbances. But the effects of these "open-loop" variations can be eliminated or at least considerably reduced by the introduction of Feedback.

A feedback system is one in which the output signal is sampled and then fed back to the input to form an error signal that drives the system.

feedback is comprised of a sub-circuit that allows a fraction of the output signal from a system to modify the effective input signal in such a way as to produce a response that can differ substantially from the response produced in the absence of such feedback.

Feedback Systems are very useful and widely used in amplifier circuits, oscillators, process control systems as well as other types of electronic systems. But for feedback to be an effective tool it must be controlled as an uncontrolled system will either oscillate or fail to function. The basic model of a feedback system is given as:

Feedback System Block Diagram Model

This basic feedback loop of sensing, controlling and actuation is the main concept behind a feedback control system and there are several good reasons why feedback is applied and used in electronic circuits:

- Circuit characteristics such as the systems gain and response can be precisely controlled.

- Circuit characteristics can be made independent of operating conditions such as supply voltages or temperature variations.

- Signal distortion due to the non-linear nature of the components used can be greatly reduced.

- The Frequency Response, Gain and Bandwidth of a circuit or system can be easily controlled to within tight limits.

Whilst there are many different types of control systems, there are just two main types of feedback control namely: Negative Feedback and Positive Feedback.

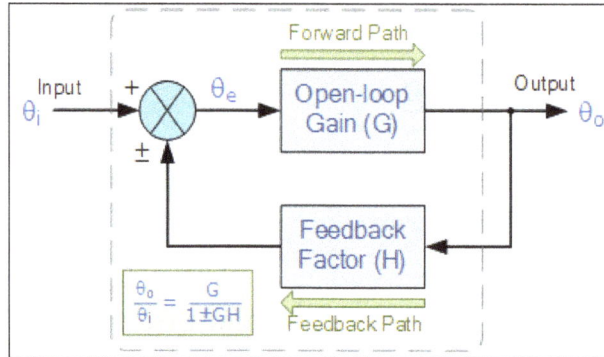

Positive Feedback Systems

In a "positive feedback control system", the set point and output values are added together by the controller as the feedback is "in-phase" with the input. The effect of positive (or regenerative) feedback is to "increase" the systems gain, i.e, the overall gain with positive feedback applied will be greater than the gain without feedback. For example, if someone praises you or gives you positive feedback about something, you feel happy about yourself and are full of energy, you feel more positive.

However, in electronic and control systems to much praise and positive feedback can increase the systems gain far too much which would give rise to oscillatory circuit responses as it increases the magnitude of the effective input signal.

An example of a positive feedback systems could be an electronic amplifier based on an operational amplifier, or op-amp as shown.

Positive Feedback System

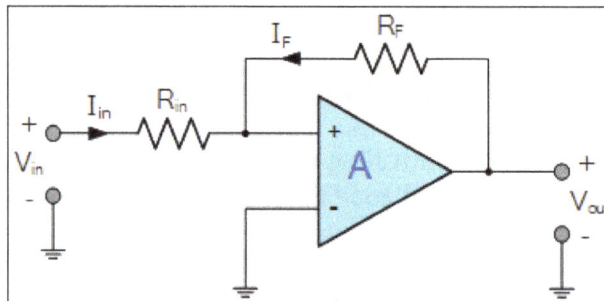

Positive feedback control of the op-amp is achieved by applying a small part of the output voltage signal at Vout back to the non-inverting (+) input terminal via the feedback resistor, R_F.

If the input voltage Vin is positive, the op-amp amplifies this positive signal and the output becomes more positive. Some of this output voltage is returned back to the input by the feedback network.

Thus the input voltage becomes more positive, causing an even larger output voltage and so on. Eventually the output becomes saturated at its positive supply rail.

Likewise, if the input voltage Vin is negative, the reverse happens and the op-amp saturates at its negative supply rail. Then we can see that positive feedback does not allow the circuit to function as an amplifier as the output voltage quickly saturates to one supply rail or the other, because with positive feedback loops "more leads to more" and "less leads to less".

Then if the loop gain is positive for any system the transfer function will be: Av = G / (1 – GH). Note that if GH = 1 the system gain Av = infinity and the circuit will start to self-oscillate, after which no input signal is needed to maintain oscillations, which is useful if you want to make an oscillator.

Although often considered undesirable, this behaviour is used in electronics to obtain a very fast switching response to a condition or signal. One example of the use of positive feedback is hysteresis in which a logic device or system maintains a given state until some input crosses a preset threshold. This type of behaviour is called "bi-stability" and is often associated with logic gates and digital switching devices such as multivibrators.

We have seen that positive or regenerative feedback increases the gain and the possibility of instability in a system which may lead to self-oscillation and as such, positive feedback is widely used in oscillatory circuits such as *Oscillators* and *Timing* circuits.

Negative Feedback Systems

In a "negative feedback control system", the set point and output values are subtracted from each other as the feedback is "out-of-phase" with the original input. The effect of negative (or degenerative) feedback is to "reduce" the gain. For example, if someone criticises you or gives you negative feedback about something, you feel unhappy about yourself and therefore lack energy, you feel less positive.

Because negative feedback produces stable circuit responses, improves stability and increases the operating bandwidth of a given system, the majority of all control and feedback systems is degenerative reducing the effects of the gain.

An example of a negative feedback system is an electronic amplifier based on an operational amplifier as shown.

Negative feedback control of the amplifier is achieved by applying a small part of the output voltage signal at Vout back to the inverting (–) input terminal via the feedback resistor, Rf.

If the input voltage Vin is positive, the op-amp amplifies this positive signal, but because its connected to the inverting input of the amplifier, and the output becomes more negative. Some of this output voltage is returned back to the input by the feedback network of Rf.

Thus the input voltage is reduced by the negative feedback signal, causing an even smaller output voltage and so on. Eventually the output will settle down and become stabilised at a value determined by the gain ratio of Rf ÷ Rin.

Likewise, if the input voltage Vin is negative, the reverse happens and the op-amps output becomes positive (inverted) which adds to the negative input signal. Then we can see that negative feedback allows the circuit to function as an amplifier, so long as the output is within the saturation limits.

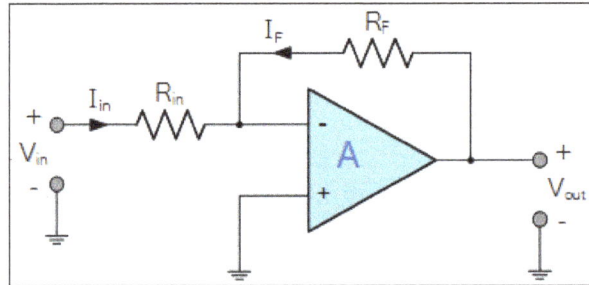

So we can see that the output voltage is stabilised and controlled by the feedback, because with negative feedback loops "more leads to less" and "less leads to more".

Then if the loop gain is positive for any system the transfer function will be:

$$Av = G / (1 + GH).$$

The use of negative feedback in amplifier and process control systems is widespread because as a rule negative feedback systems are more stable than positive feedback systems, and a negative feedback system is said to be stable if it does not oscillate by itself at any frequency except for a given circuit condition.

Another advantage is that negative feedback also makes control systems more immune to random variations in component values and inputs. Of course nothing is for free, so it must be used with caution as negative feedback significantly modifies the operating characteristics of a given system.

Classification of Feedback Systems

Thus far we have seen the way in which the output signal is "fed back" to the input terminal, and for feedback systems this can be of either, Positive Feedback or Negative Feedback. But the manner in which the output signal is measured and introduced into the input circuit can be very different leading to four basic classifications of feedback.

Based on the input quantity being amplified, and on the desired output condition, the input and output variables can be modelled as either a voltage or a current. As a result, there are four basic classifications of single-loop feedback system in which the output signal is fed back to the input and these are:

- Series-Shunt Configuration: Voltage in and Voltage out or Voltage Controlled Voltage Source (VCVS).

- Shunt-Shunt Configuration: Current in and Voltage out or Current Controlled Voltage Source (CCVS).

- Series-Series Configuration: Voltage in and Current out or Voltage Controlled Current Source (VCCS).

- Shunt-Series Configuration: Current in and Current out or Current Controlled Current Source (CCCS).

These names come from the way that the feedback network connects between the input and output stages as shown.

Series-Shunt Feedback Systems

Series-Shunt Feedback, also known as *series voltage feedback*, operates as a voltage-voltage controlled feedback system. The error voltage fed back from the feedback network is in *series* with the input. The voltage which is fed back from the output being proportional to the output voltage, Vo as it is parallel, or shunt connected.

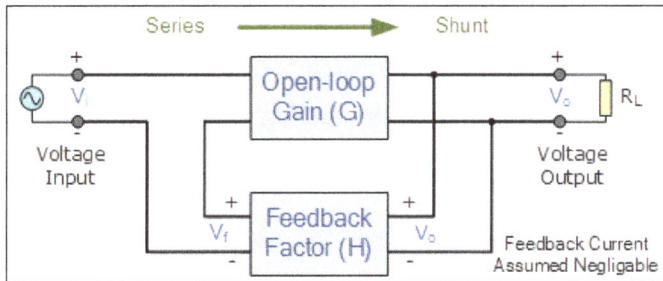

Series-Shunt Feedback System.

For the series-shunt connection, the configuration is defined as the output voltage, Vout to the input voltage, Vin. Most inverting and non-inverting operational amplifier circuits operate with series-shunt feedback producing what is known as a "voltage amplifier". As a voltage amplifier the ideal input resistance, Rin is very large, and the ideal output resistance, Rout is very small.

Then the "series-shunt feedback configuration" works as a true voltage amplifier as the input signal is a voltage and the output signal is a voltage, so the transfer gain is given as: Av = Vout ÷ Vin. Note that this quantity is dimensionless as its units are volts/volts.

Shunt-Series Feedback Systems

Shunt-Series Feedback, also known as shunt current feedback, operates as a current-current controlled feedback system. The feedback signal is proportional to the output current, Io flowing in the load. The feedback signal is fed back in parallel or shunt with the input as shown.

For the shunt-series connection, the configuration is defined as the output current, Iout to the input current, Iin. In the shunt-series feedback configuration the signal fed back is in parallel with the input signal and as such its the currents, not the voltages that add.

This parallel shunt feedback connection will not normally affect the voltage gain of the system, since for a voltage output a voltage input is required. Also, the series connection at the output increases output resistance, Rout while the shunt connection at the input decreases the input resistance, Rin.

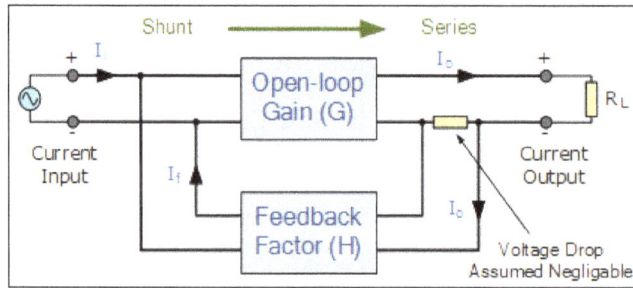

Shunt-Series Feedback System.

Then the "shunt-series feedback configuration" works as a true current amplifier as the input signal is a current and the output signal is a current, so the transfer gain is given as: Ai = Iout ÷ Iin. Note that this quantity is dimensionless as its units are amperes/amperes.

Series-Series Feedback Systems

Series-Series Feedback Systems, also known as series current feedback, operates as a voltage-current controlled feedback system. In the series current configuration the feedback error signal is in series with the input and is proportional to the load current, Iout. Actually, this type of feedback converts the current signal into a voltage which is actually fed back and it is this voltage which is subtracted from the input.

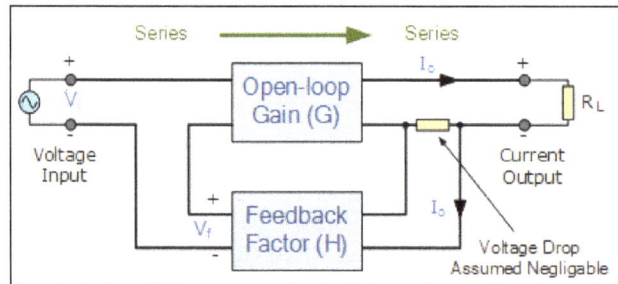

Series-Series Feedback System.

For the series-series connection, the configuration is defined as the output current, Iout to the input voltage, Vin. Because the output current, Iout of the series connection is fed back as a voltage, this increases both the input and output impedances of the system. Therefore, the circuit works best as a transconductance amplifier with the ideal input resistance, Rin being very large, and the ideal output resistance, Rout is also very large.

Then the "series-series feedback configuration" functions as transconductance type amplifier system as the input signal is a voltage and the output signal is a current. then for a series-series feedback circuit the transfer gain is given as: Gm = Iout ÷ Vin.

Shunt-Shunt Feedback Systems

Shunt-Shunt Feedback Systems, also known as shunt voltage feedback, operates as a current-voltage controlled feedback system. In the shunt-shunt feedback configuration the signal fed back is in parallel with the input signal. The output voltage is sensed and the current is subtracted from the input current in shunt, and as such its the currents, not the voltages that subtract.

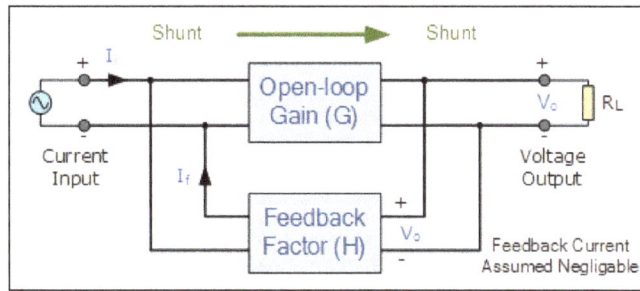

Shunt-Shunt Feedback System.

For the shunt-shunt connection, the configuration is defined as the output voltage, Vout to the input current, Iin. As the output voltage is fed back as a current to a current-driven input port, the shunt connections at both the input and output terminals reduce the input and output impedance. therefore the system works best as a transresistance system with the ideal input resistance, Rin being very small, and the ideal output resistance, Rout also being very small.

Then the shunt voltage configuration works as transresistance type voltage amplifier as the input signal is a current and the output signal is a voltage, so the transfer gain is given as: Rm = Vout ÷ Iin.

Feedback Systems Summary

We have seen that a Feedback System is one in which the output signal is sampled and then fed back to the input to form an error signal that drives the system, and depending on the type of feedback used, the feedback signal which is mixed with the systems input signal, can be either a voltage or a current.

Feedback will always change the performance of a system and feedback arrangements can be either positive (regenerative) or negative (degenerative) type feedback systems. If the feedback loop around the system produces a loop-gain which is negative, the feedback is said to be negative or degenerative with the main effect of the negative feedback is in reducing the systems gain.

If however the gain around the loop is positive, the system is said to have positive feedback or regenerative feedback. The effect of positive feedback is to increase the gain which can cause a system to become unstable and oscillate especially if GH = -1.

We have also seen that block-diagrams can be used to demonstrate the various types of feedback systems. In the block diagrams above, the input and output variables can be modelled as either a voltage or a current and as such there are four combinations of inputs and outputs that represent the possible types of feedback, namely: Series Voltage Feedback, Shunt Voltage Feedback, Series Current Feedback and Shunt Current Feedback.

The names for these different types of feedback systems are derived from the way that the feedback network connects between the input and output stages either in parallel (shunt) or series.

Proportional Control

Proportional control, in engineering and process control, is a type of linear feedback control system in which a correction is applied to the controlled variable which is proportional to the difference between the desired value (set point, SP) and the measured value (process value, PV). Two classic mechanical examples are the toilet bowl float proportioning valve and the fly-ball governor.

The proportional control concept is more complex than an on–off control system like a bi-metallic domestic thermostat, but simpler than a proportional–integral–derivative (PID) control system used in something like an automobile cruise control. On–off control will work where the overall system has a relatively long response time, but can result in instability if the system being controlled has a rapid response time. Proportional control overcomes this by modulating the output to the controlling device, such as a control valve at a level which avoids instability, but applies correction as fast as practicable by applying the optimum quantity of proportional gain.

A drawback of proportional control is that it cannot eliminate the residual SP − PV error in processes with compensation e.g. temperature control, as it requires an error to generate a proportional output. To overcome this the PI controller was devised, which uses a proportional term (P) to remove the gross error, and an integral term (I) to eliminate the residual offset error by integrating the error over time to produce an "I" component for the controller output.

Theory

In the proportional control algorithm, the controller output is proportional to the error signal, which is the difference between the setpoint and the process variable. In other words, the output of a proportional controller is the multiplication product of the error signal and the proportional gain.

This can be mathematically expressed as,

$$P_{\text{out}} = K_p e(t) + p0$$

where,

- $p0$: Controller output with zero error.

- P_{out} : Output of the proportional controller.

- K_p : Proportional gain.

- $e(t)$: Instantaneous process error at time t. $e(t) = SP - PV$.

- SP : Set point.

- PV : Process variable.

Constraints: In a real plant, actuators have physical limitations that can be expressed as constraints on P_{out}. For example, P_{out} may be bounded between −1 and +1 if those are the maximum output limits.

Qualifications: It is preferable to express K_p as a unitless number. To do this, we can express $e(t)$ as a ratio with the span of the instrument. This span is in the same units as error (e.g. C degrees) so the ratio has no units.

Development of Control Block Diagrams

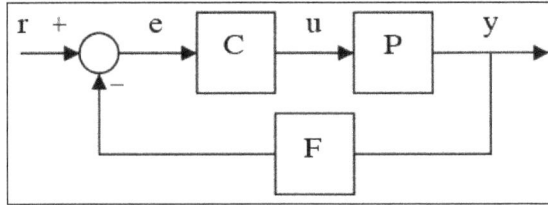

Simple feedback control loop.

Proportional control dictates $g_c = k_c$. From the block diagram shown, assume that r, the setpoint, is the flowrate into a tank and e is *error*, which is the difference between setpoint and measured process output. g_p, is process transfer function; the input into the block is flow rate and output is tank level.

The output as a function of the setpoint, r, is known as the *closed-loop transfer function*. $g_{cl} = \dfrac{g_p g_c}{1 + g_p g_c}$, If the poles of g_{cl}, are stable, then the closed-loop system is stable.

First-order Process

For a first-order process, a general transfer function is $g_p = \dfrac{k_p}{\tau_p s + 1}$. Combining this with the closed-loop transfer function above returns $g_{CL} = \dfrac{\dfrac{k_p k_c}{\tau_p s + 1}}{1 + \dfrac{k_p k_c}{\tau_p s + 1}}$. Simplifying this equation results in

$g_{CL} = \dfrac{k_{CL}}{\tau_{CL} s + 1}$ where $k_{CL} = \dfrac{k_p k_c}{1 + k_p k_c}$ and $\tau_{CL} = \dfrac{\tau_p}{1 + k_p k_c}$. For stability in this system, $\tau_{CL} > 0$; there-

fore, τ_p must be a positive number, and $k_p k_c > -1$ (standard practice is to make sure that $k_p k_c > 0$). Introducing a step change to the system gives the output response of $y(s) = g_{CL} \times \dfrac{\Delta R}{s}$.

Using the final-value theorem,

$$\lim_{t \to \infty} y(t) = \lim_{s \searrow 0} \left(s \times \frac{k_{CL}}{\tau_{CL} s + 1} \times \frac{\Delta R}{s} \right) = k_{CL} \times \Delta R = y(t)\big|_{t = \infty}$$

which shows that there will always be an offset in the system.

Integrating Process

For an integrating process, a general transfer function is $g_p = \dfrac{1}{s(s+1)}$, which, when combined

with the closed-loop transfer function, becomes $g_{CL} = \dfrac{k_c}{s(s+1) + k_c}$.

Introducing a step change to the system gives the output response of $y(s) = g_{CL} \times \dfrac{\Delta R}{s}$.

Using the final-value theorem,

$$\lim_{t \to \infty} y(t) = \lim_{s \searrow 0}\left(s \times \frac{k_c}{s(s+1)+k_c} \times \frac{\Delta R}{s} \right) = \Delta R = y(t) \big|_{t=\infty}$$

meaning there is no offset in this system. This is the only process that will not have any offset when using a proportional controller.

Offset Error

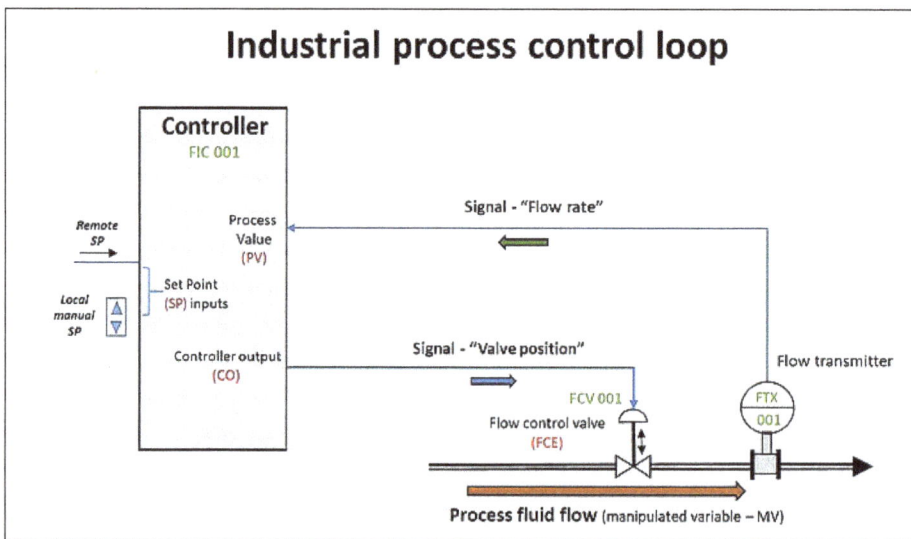

Flow control loop: If only a proportional controller, then there's always an offset between SP and PV.

Proportional control cannot eliminate the offset error, which is the difference between the desired value and the actual value, SP – PV error, as it requires an error to generate an output. When a disturbance (deviation from existing state) occurs in the process value being controlled, any corrective control action, based purely on Proportional Control, will always leave out the error between the next steady state and the desired setpoint, and result in a residual error called the offset error. This error will increase as greater process demand is put on the system, or by increasing the set point.

Consider an object suspended by a spring as a simple proportional control. The spring will attempt to maintain the object in a certain location despite disturbances which may temporarily displace it. Hooke's law tells us that the spring applies a corrective force that is proportional to the object's displacement. While this will tend to hold the object in a particular location, the absolute resting location of the object will vary if its mass is changed. This difference in resting location is the offset error.

Imagine the same spring and object in a weightless environment. In this case, the spring will tend to hold the object in the same location regardless of its mass. There is no offset error in this case because the proportional action is not working against anything in the steady state.

Proportional Band

The proportional band is the band of controller output over which the final control element (a control valve, for instance) will move from one extreme to another. Mathematically, it can be expressed as:

$$PB = \frac{100}{K_p}$$

So if K_p, the proportional gain, is very high, the proportional band is very small, which means that the band of controller output over which the final control element will go from minimum to maximum (or vice versa) is very small. This is the case with on–off controllers, where K_p is very high and hence, for even a small error, the controller output is driven from one extreme to another.

Advantages

The clear advantage of proportional over on–off control can be demonstrated by car speed control. An analogy to on–off control is driving a car by applying either full power or no power and varying the duty cycle, to control speed. The power would be on until the target speed is reached, and then the power would be removed, so the car reduces speed. When the speed falls below the target, with a certain hysteresis, full power would again be applied. It can be seen that this would obviously result in poor control and large variations in speed. The more powerful the engine; the greater the instability, the heavier the car; the greater the stability. Stability may be expressed as correlating to the power-to-weight ratio of the vehicle.

In proportional control, the power output is always proportional to the (actual versus target speed) error. If the car is at target speed and the speed increases slightly due to a falling gradient, the power is reduced slightly, or in proportion to the change in error, so that the car reduces speed gradually and reaches the new target point with very little, if any, "overshoot", which is much smoother control than on–off control. In practice, PID controllers are used for this and the large number of control processes that require more response control than proportional alone.

Bang–bang Control

In control theory, a bang–bang controller (2 step or on–off controller), also known as a hysteresis controller, is a feedback controller that switches abruptly between two states. These controllers may be realized in terms of any element that provides hysteresis. They are often used to control a plant that accepts a binary input, for example a furnace that is either completely on or completely off. Most common residential thermostats are bang–bang controllers. The Heaviside step function in its discrete form is an example of a bang–bang control signal. Due to the discontinuous control signal, systems that include bang–bang controllers are variable structure systems, and bang–bang controllers are thus variable structure controllers.

A water heater that maintains desired temperature by turning the applied power on and off (as opposed to continuously varying electrical voltage or current) based on temperature feedback is an example application of bang–bang control. Although the applied power switches from one discrete state to another, the water temperature will remain relatively constant due to the slow nature of temperature changes in materials. Hence, the regulated temperature is like a sliding mode of the variable structure system setup by the bang–bang controller.

Bang–bang solutions in Optimal Control

In optimal control problems, it is sometimes the case that a control is restricted to be between a lower and an upper bound. If the optimal control switches from one extreme to the other (i.e., is strictly never in between the bounds), then that control is referred to as a bang-bang solution.

Bang–bang controls frequently arise in minimum-time problems. For example, if it is desired to stop a car in the shortest possible time at a certain position sufficiently far ahead of the car, the solution is to apply maximum acceleration until the unique *switching point*, and then apply maximum braking to come to rest exactly at the desired position.

A familiar everyday example is bringing water to a boil in the shortest time, which is achieved by applying full heat, then turning it off when the water reaches a boil. A closed-loop household example is most thermostats, wherein the heating element or air conditioning compressor is either running or not, depending upon whether the measured temperature is above or below the setpoint.

Bang–bang solutions also arise when the Hamiltonian is linear in the control variable; application of Pontryagin's minimum or maximum principle will then lead to pushing the control to its upper or lower bound depending on the sign of the coefficient of u in the Hamiltonian.

In summary, bang–bang controls are actually optimal controls in some cases, although they are also often implemented because of simplicity or convenience.

Practical Implications of Bang-bang Control

Mathematically or within a computing context there may be no problems, but the physical realization of bang-bang control systems gives rise to several complications.

First, depending on the width of the hysteresis gap and inertia in the process, there will be an oscillating error signal around the desired set point value (e.g., temperature), often saw-tooth shaped. Room temperature may become uncomfortable just before the next switch 'ON' event. Alternatively, a narrow hysteresis gap will lead to frequent on/off switching, which is undesirable for, e.g., an electrically ignited gas heater.

Second, the onset of the step function may entail, for example, a high electrical current and/or sudden heating and expansion of metal vessels, ultimately leading to metal fatigue or other wear-and-tear effects. Where possible, continuous control, such as in PID control will avoid problems caused by the brisk physical system state transitions that are the consequence of bang-bang control.

ROOT LOCUS

The root locus of a feedback system is the graphical representation in the complex s-plane of the possible locations of its closed-loop poles for varying values of a certain system parameter. The points that are part of the root locus satisfy the angle condition. The value of the parameter for a certain point of the root locus can be obtained using the magnitude condition.

Suppose there is a feedback system with input signal $X(s)$ and output signal $Y(s)$. The forward path transfer function is $G(s)$; the feedback path transfer function is $H(s)$.

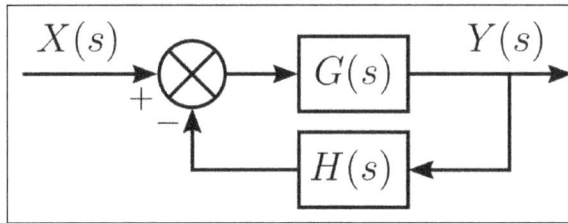

For this system, the closed-loop transfer function is given by,

$$T(s) = \frac{Y(s)}{X(s)} = \frac{G(s)}{1+G(s)H(s)}$$

Thus, the closed-loop poles of the closed-loop transfer function are the roots of the characteristic equation $G(s)H(s) = -1$. The roots of this equation may be found wherever $1+G(s)H(s) = 0$.

In systems without pure delay, the product $G(s)H(s)$ is a rational polynomial function and may be expressed as,

$$G(s)H(s) = K\frac{(s+z_1)(s+z_2)\cdots(s+z_m)}{(s+p_1)(s+p_2)\cdots(s+p_n)}$$

where $-z_i$ are the m zeros, $-p_i$ are the n poles, and K is a scalar gain. Typically, a root locus diagram will indicate the transfer function's pole locations for varying values of the parameter K. A root locus plot will be all those points in the s-plane where $G(s)H(s) = -1$ for any value of K.

The factoring of K and the use of simple monomials means the evaluation of the rational polynomial can be done with vector techniques that add or subtract angles and multiply or divide magnitudes. The vector formulation arises from the fact that each monomial term $(s-a)$ in the factored

$G(s)H(s)$ represents the vector from a to s in the s-plane. The polynomial can be evaluated by considering the magnitudes and angles of each of these vectors.

According to vector mathematics, the angle of the result of the rational polynomial is the sum of all the angles in the numerator minus the sum of all the angles in the denominator. So to test whether a point in the s-plane is on the root locus, only the angles to all the open loop poles and zeros need be considered. This is known as the angle condition.

Similarly, the magnitude of the result of the rational polynomial is the product of all the magnitudes in the numerator divided by the product of all the magnitudes in the denominator. It turns out that the calculation of the magnitude is not needed to determine if a point in the s-plane is part of the root locus because K varies and can take an arbitrary real value. For each point of the root locus a value of K can be calculated. This is known as the magnitude condition.

The root locus only gives the location of closed loop poles as the gain K is varied. The value of K does not affect the location of the zeros. The open-loop zeros are the same as the closed-loop zeros.

Angle Condition

A point s of the complex s-plane satisfies the angle condition if,

$$\angle(G(s)H(s)) = \pi$$

which is the same as saying that,

$$\sum_{i=1}^{m} \angle(s + z_i) - \sum_{i=1}^{n} \angle(s + p_i) = \pi$$

that is, the sum of the angles from the open-loop zeros to the point s minus the angles from the open-loop poles to the point s has to be equal to π, or 180 degrees.

Magnitude Condition

A value of K satisfies the magnitude condition for a given s point of the root locus if,

$$|G(s)H(s)| = 1$$

which is the same as saying that,

$$K \frac{|s + z_1||s + z_2|\cdots|s + z_m|}{|s + p_1||s + p_2|\cdots|s + p_n|} = 1.$$

In control theory and stability theory, root locus analysis is a graphical method for examining how the roots of a system change with variation of a certain system parameter, commonly a gain within a feedback system. This is a technique used as a stability criterion in the field of classical control theory developed by Walter R. Evans which can determine stability of the system. The root locus plots the poles of the closed loop transfer function in the complex s-plane as a function of a gain parameter.

A graphical method that uses a special protractor called a "Spirule" was once used to determine angles and draw the root loci.

Uses

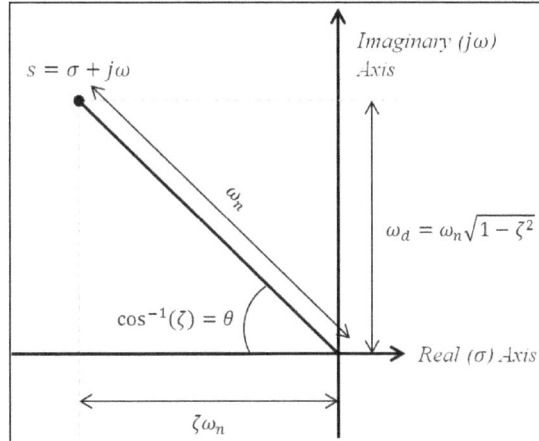

Effect of pole location on a second order system's natural frequency and damping ratio. Note that this pole's complex conjugate (which necessarily exists since this pole has a nonzero imaginary component) is not shown.

In addition to determining the stability of the system, the root locus can be used to design the damping ratio (ζ) and natural frequency (ω_n) of a feedback system. Lines of constant damping ratio can be drawn radially from the origin and lines of constant natural frequency can be drawn as arccosine whose center points coincide with the origin. By selecting a point along the root locus that coincides with a desired damping ratio and natural frequency, a gain K can be calculated and implemented in the controller. More elaborate techniques of controller design using the root locus are available in most control textbooks: for instance, lag, lead, PI, PD and PID controllers can be designed approximately with this technique.

The definition of the damping ratio and natural frequency presumes that the overall feedback system is well approximated by a second order system; i.e. the system has a dominant pair of poles. This is often not the case, so it is good practice to simulate the final design to check if the project goals are satisfied.

Sketching Root Locus

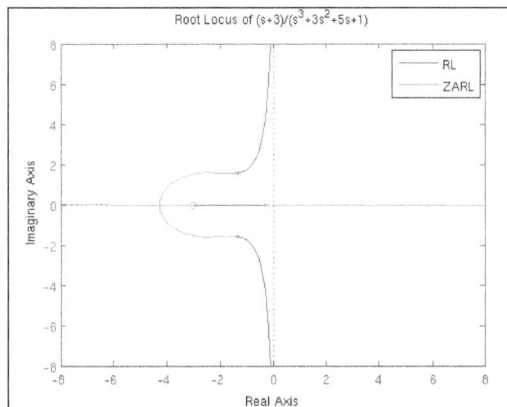

RL = root locus; ZARL = zero angle root locus.

Using a few basic rules, the root locus method can plot the overall shape of the path (locus) traversed by the roots as the value of K varies. The plot of the root locus then gives an idea of the stability and dynamics of this feedback system for different values of K. The rules are the following:

- Mark open-loop poles and zeros.

- Mark real axis portion to the left of an odd number of poles and zeros.

- Find asymptotes.

Let P be the number of poles and Z be the number of zeros:

$P - Z$ = number of asymptotes

The asymptotes intersect the real axis at α (which is called the centroid) and depart at angle ϕ given by:

$$\phi_l = \frac{180^\circ + (l-1)360^\circ}{P-Z}, l = 1, 2, \ldots, P-Z$$

$$\alpha = \frac{\sum_P - \sum_z}{P-Z}$$

where $\sum_P$ is the sum of all the locations of the poles, and $\sum_Z$ is the sum of all the locations of the explicit zeros.

- Phase condition on test point to find angle of departure.

- Compute breakaway/break-in points.

The breakaway points are located at the roots of the following equation:

$$\frac{dG(s)H(s)}{ds} = 0 \text{ or } \frac{d\overline{GH}(z)}{dz} = 0$$

Once you solve for z, the real roots give you the breakaway/reentry points. Complex roots correspond to a lack of breakaway/reentry.

z-plane versus s-plane

The root locus method can also be used for the analysis of sampled data systems by computing the root locus in the z-plane, the discrete counterpart of the s-plane. The equation $z = e^{sT}$ maps continuous s-plane poles (not zeros) into the z-domain, where T is the sampling period. The stable, left half s-plane maps into the interior of the unit circle of the z-plane, with the s-plane origin equating to $|z| = 1$ (because $e^0 = 1$). A diagonal line of constant damping in the s-plane maps around a spiral from (1,0) in the z plane as it curves in toward the origin. Note also that the Nyquist aliasing criteria is expressed graphically in the z-plane by the x-axis, where $\omega nT = \pi$. The line of constant damping just described spirals in indefinitely but in sampled data systems, frequency content is aliased down to lower frequencies by integral multiples of the Nyquist frequency. That is, the sampled response appears as a lower frequency and better damped as well since the root in the z-plane maps equally well to the first loop of a different, better damped spiral curve of constant damping. Many other interesting and relevant mapping properties can be described, not least that z-plane controllers, having the property that they may be directly implemented from the z-plane transfer function (zero/pole ratio of polynomials), can be imagined graphically on a z-plane plot of the open loop transfer function, and immediately analyzed utilizing root locus.

Since root locus is a graphical angle technique, root locus rules work the same in the z and s planes.

The idea of a root locus can be applied to many systems where a single parameter K is varied. For example, it is useful to sweep any system parameter for which the exact value is uncertain in order to determine its behavior.

OPEN-LOOP CONTROLLER

In an open-loop controller, also called a non-feedback controller, the control action from the controller is independent of the "process output", which is the process variable that is being controlled. It does not use feedback to determine if its output has achieved the desired goal of the input command or process "set point".

There are a large number of open loop controls, such as on/off switching of valves, machinery, lights, motors or heaters, where the control result is known to be approximately sufficient under normal conditions without the need for feedback. The advantage of using open loop control in these cases is the reduction in component count and complexity. However, an open-loop system cannot correct any errors that it makes or correct for outside disturbances, and cannot engage in machine learning.

Open-loop and Closed-loop (Feedback) Control

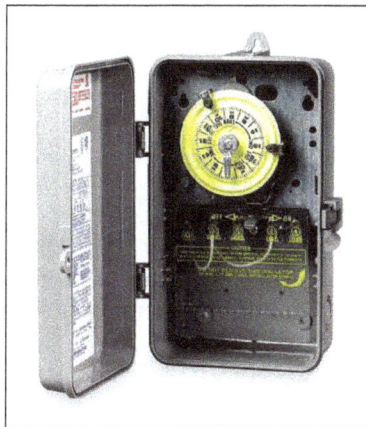

An electromechanical timer, normally used for open-loop control based purely on a timing sequence, with no feedback from the process.

Fundamentally, there are two types of control loop: open loop (feedforward) control, and closed loop (feedback) control.

In open loop control, the control action from the controller is independent of the "process output" (or "controlled process variable"). A good example of this is a central heating boiler controlled only by a timer, so that heat is applied for a constant time, regardless of the temperature of the building. The control action is the switching on/off of the boiler, but the controlled variable should be the building temperature, but is not as this is open-loop control of the boiler, which does not give closed-loop control of the temperature.

In closed loop control, the control action from the controller is dependent on the process output. In the case of the boiler analogy this would include a thermostat to monitor the building temperature, and thereby feed back a signal to ensure the controller maintains the building at the temperature set on the thermostat. A closed loop controller therefore has a feedback loop which ensures the controller exerts a control action to give a process output the same as the "reference input" or "set point". For this reason, closed loop controllers are also called feedback controllers.

The definition of a closed loop control system according to the British Standard Institution is "a control system possessing monitoring feedback, the deviation signal formed as a result of this feedback being used to control the action of a final control element in such a way as to tend to reduce the deviation to zero."

Applications

An open-loop controller is often used in simple processes because of its simplicity and low cost, especially in systems where feedback is not critical. A typical example would be an older model domestic clothes dryer, for which the length of time is entirely dependent on the judgement of the human operator, with no automatic feedback of the dryness of the clothes.

Electric clothes dryer, which is open loop controlled by running the dryer
for a set time, regardless of clothes dryness.

For example, an irrigation sprinkler system, programmed to turn on at set times could be an example of an open-loop system if it does not measure soil moisture as a form of feedback. Even if rain is pouring down on the lawn, the sprinkler system would activate on schedule, wasting water.

Another example is a Stepper motors used for control of position. Sending it a stream of electrical pulses causes it to rotate by exactly that many steps, hence the name. If the motor was always assumed to perform each movement correctly, without positional feedback, it would be open loop control. However, if there is a position encoder, or sensors to indicate the "start" or finish positions, then that is closed-loop control, such as in many inkjet printers. The drawback of open-loop control of steppers is that if the machine load is too high, or the motor attempts to move too quickly, then steps may be skipped. The controller has no means of detecting this and so the machine continues to run slightly out of adjustment until reset. For this reason, more complex robots and machine tools instead use servomotors rather than stepper motors, which incorporate encoders and closed-loop controllers.

However, open-loop control is very useful and economic for well-defined systems where the relationship between input and the resultant state can be reliably modeled by a mathematical formula. For example, determining the voltage to be fed to an electric motor that drives a constant load, in order to achieve a desired speed would be a good application. But if the load were not predictable and became excessive, the motor's speed might vary as a function of the load not just the voltage, and an open-loop controller would be insufficient to ensure repeatable control of the velocity.

An example of this is a conveyor system that is required to travel at a constant speed. For a constant voltage, the conveyor will move at a different speed depending on the load on the motor (represented here by the weight of objects on the conveyor). In order for the conveyor to run at a constant speed, the voltage of the motor must be adjusted depending on the load. In this case, a closed-loop control system would be necessary.

Thus there are a large number of open loop controls, such as switching valves, lights, motors or heaters on and off, where the result is known to be approximately sufficient without the need for feedback.

Feedback Control

A feed back control system, such as a PID controller, can be improved by combining the feedback (or closed-loop) control of a PID controller with feed-forward (or open-loop) control. Knowledge about the system (such as the desired acceleration and inertia) can be fed forward and combined with the PID output to improve the overall system performance. The feed-forward value alone can often provide the major portion of the controller output. The PID controller primarily has to compensate whatever difference or *error* remains between the setpoint (SP) and the system response to the open loop control. Since the feed-forward output is not affected by the process feedback, it can never cause the control system to oscillate, thus improving the system response without affecting stability. Feed forward can be based on the setpoint and on extra measured disturbances. Setpoint weighting is a simple form of feed forward.

For example, in most motion control systems, in order to accelerate a mechanical load under control, more force is required from the actuator. If a velocity loop PID controller is being used to control the speed of the load and command the force being applied by the actuator, then it is beneficial to take the desired instantaneous acceleration, scale that value appropriately and add it to the output of the PID velocity loop controller. This means that whenever the load is being accelerated or decelerated, a proportional amount of force is commanded from the actuator regardless of the feedback value. The PID loop in this situation uses the feedback information to change the combined output to reduce the remaining difference between the process setpoint and the feedback value. Working together, the combined open-loop feed-forward controller and closed-loop PID controller can provide a more responsive control system in some situations.

DIGITAL CONTROL

Digital control is a branch of control theory that uses digital computers to act as system controllers. Depending on the requirements, a digital control system can take the form of a microcontroller to an ASIC to a standard desktop computer. Since a digital computer is a discrete system, the Laplace transform is replaced with the Z-transform. Also since a digital computer has finite precision extra care is needed to ensure the error in coefficients, A/D conversion, D/A conversion, etc. are not producing undesired or unplanned effects.

The application of digital control can readily be understood in the use of feedback. Since the creation of the first digital computer in the early 1940s the price of digital computers has dropped considerably, which has made them key pieces to control systems for several reasons:

- Inexpensive: under $5 for many microcontrollers.

- Flexible: easy to configure and reconfigure through software.

- Scalable: programs can scale to the limits of the memory or storage space without extra cost.

- Adaptable: parameters of the program can change with time.

- Static operation: digital computers are much less prone to environmental conditions than capacitors, inductors, etc.

Digital Controller Implementation

A digital controller is usually cascaded with the plant in a feedback system. The rest of the system can either be digital or analog.

Typically, a digital controller requires:

- A/D conversion to convert analog inputs to machine readable (digital) format.

- D/A conversion to convert digital outputs to a form that can be input to a plant (analog).

- A program that relates the outputs to the inputs.

Output Program

- Outputs from the digital controller are functions of current and past input samples, as well as past output samples - this can be implemented by storing relevant values of input and output in registers. The output can then be formed by a weighted sum of these stored values.

The programs can take numerous forms and perform many functions:

- A digital filter for low-pass filtering.

- A state space model of a system to act as a state observer.

- A telemetry system.

Stability

Although a controller may be stable when implemented as an analog controller, it could be unstable when implemented as a digital controller due to a large sampling interval. During sampling the aliasing modifies the cutoff parameters. Thus the sample rate characterizes the transient response and stability of the compensated system, and must update the values at the controller input often enough so as to not cause instability.

When substituting the frequency into the z operator, regular stability criteria still apply to discrete control systems. Nyquist criteria apply to z-domain transfer functions as well as being general for complex valued functions. Bode stability criteria apply similarly. Jury criterion determines the discrete system stability about its characteristic polynomial.

Design of Digital Controller in s-domain

The digital controller can also be designed in the s-domain (continuous). The Tustin transformation can transform the continuous compensator to the respective digital compensator. The digital compensator will achieve an output which approaches the output of its respective analog controller as the sampling interval is decreased.

$$s = \frac{2(z-1)}{T(z+1)}$$

Tustin Transformation Deduction

Tustin is the Padé$_{(1,1)}$ approximation of the exponential function $z = e^{sT}$:

$$z = e^{sT}$$
$$= \frac{e^{sT/2}}{e^{-sT/2}}$$
$$\approx \frac{1 + sT/2}{1 - sT/2}$$

And its inverse,

$$s = \frac{1}{T}\ln(z)$$

$$= \frac{2}{T}\left[\frac{z-1}{z+1} + \frac{1}{3}\left(\frac{z-1}{z+1}\right)^3 + \frac{1}{5}\left(\frac{z-1}{z+1}\right)^5 + \frac{1}{7}\left(\frac{z-1}{z+1}\right)^7 + \cdots\right]$$

$$\approx \frac{2}{T}\frac{z-1}{z+1}$$

$$= \frac{2}{T}\frac{1-z^{-1}}{1+z^{-1}}$$

We must never forget that the digital control theory is the technique to design strategies in discrete time, (and/or) quantized amplitude (and/or) in (binary) coded form to be implemented in computer systems (microcontrollers, microprocessors) that will control the analog (continuous in time and amplitude) dynamics of analog systems.

APPLICATIONS OF CONTROL THEORY IN BIOMEDICAL ENGINEERING

The control mechanism composes basic for maintenance of homeostasis at all levels of organization in the hierarchy of living systems. The control law of biologic system is evolved through millions of years. It is perfect to balance every physiologic state, and to maintain human life. From the aging, the function of organs and tissues is weakened little by little. The disease or accident can also impair the function of living systems. Finally, the control mechanism of physiology has broken and the unstable biologic systems could endanger the life. The goal of drug dosing and medical device assisting is to recover natural control mechanism. It is hard to completely replace natural control mechanism by artificial method. So, to study the control theory in biomedical engineering is necessary.

The objective of this article is to explore some biomedical control problem and solution to general control researcher. We wish to interest the control community in the idea of developing applications in medicine, and demonstrate to medical community that control theory had solid applications in the medical field. Besides, biomedical system provides classic examples of effectively functioning control system design. Medical topic studied regarding their control structures and behavior can benefit in other domains of control engineering practice.

Before going deeply into the biomedical control, we should understand the difference between engineering and physiological control systems. At first, the engineering control system is designed for a specific task, and we can fine-tune the controller parameter to achieve optimal result. In contrast, physiologic control systems are built for versatility, and may be capable of several different functions. Since the designer develops the engineering system, the characteristics of the model are generally known. On the other hand, the physiological system is usually unknown and difficult to analyze. Moreover, there exist serious cross-coupling among different physiological control system. As for the engineering systems, the state variables of model may be independent (linear). The controller of physiological systems, in general, is adaptive. This mean the control system may be able to change the output not only through feedback, but also by allowing the controller characteristics change. And the engineering system can only use the feedback controller to achieve the objective. About the establishing of feedback path, the feedback signal of engineering control systems is explicitly subtracted from the reference input, revealing clearly the using of negative feedback. However, negative feedback in the physiological system is embedded in the plant characteristics. Table gives a summary.

Table: Differences between engineering and physiological control systems.

	Engineering control systems	Physiological control systems
Goal	For a specific task	Built for versatility
Design	Fine tuned parameter (optimal)	Capable of serving different functions
Model	Generally known	Unknown and difficult to analyze
State variables	Independent	Cross-coupling
Controller	Feedback	Adaptive
Feedback	Explicitly	Embedded

In rehabilitation engineering, we talk about the control of powered prosthesis, functional electrical stimulation and biofeedback control. In the section of drug delivery, we discuss the model analysis and control strategy in immune system, blood pressure regulation, blood glucose control and anesthesia control. In the section 3, we also consider the control technique for developing medical device.

Rehabilitation Engineering

Rehabilitation engineering can be described as the device, method and application to meet the needs of person with disabilities in engineering method. Fource Depend on the help of assist devices; disabilities can improve the quality of living and return to original life. In the following, we discuss some rehabilitation technique, like powered prostheses, functional electrical stimulation and biofeedback control.

The Control of Powered Prostheses

The prostheses used in the limbs of the individuals whom with amputations or congenitally deficient extremities. In current technique, however, prosthesis can be considered as a tool rather than a limbs replacement. The real problem is the issue of how to interface a multifunction arm or leg to an amputee in a meaningful way. For this reason that the prosthesis is often dominated by consideration of control.

Electromyography (EMG) signal is the best way to achieve multifunctional prosthesis control. Englehart et al. have demonstrated the potential of myoelectric base powered prosthesis control design. The fuzzy approach to classify single-site electromyograph (EMG) signals for multifunctional prosthesis control is also presented in. The fuzzy approach was compared with an artificial neural network (ANN) method on four subjects, and a slightly superior classification results were obtained.

We need to develop progress toward a more natural, more effective means of myoelectric control by providing high accuracy, low response time, and an intuitive control interface to the user.

Functional Electrical Stimulation

There are quite difference between powered prosthesis and functional electrical stimulation, powered prosthesis use body signal to control external assist device but function electrical stimulation use external signal to control body. Absence of basic functions such as grasping, walking, breathing, and bladder voiding often renders such subjects dependent on the assistance of others for daily living activities. This cost the government lots of money and efforts. Functional electrical stimulation (FES) is the use of voltage pulses to induce skeletal muscle contraction and consequently joint movements. By precisely controlling electrical stimulation, we can change the body postural and accomplish many actions, which include balance, standing and walking etc. FES has been successfully adopted for stroke rehabilitation, spinal cord injuries and many limbs motor dysfunction in clinic. Figure shows the block diagram of functional electrical stimulation control.

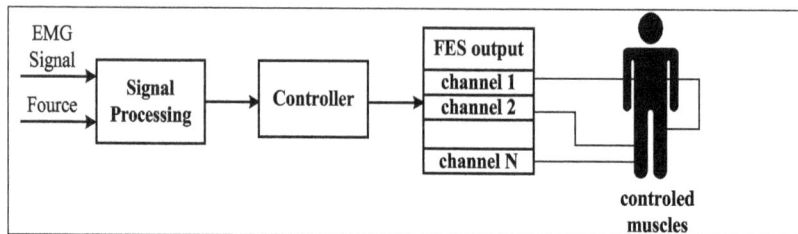

Functional electrical stimulation control block diagram.

Balance Control

Balance is the most import, because stand, walk, and many activities all rely on this ability. The frequency of falls in elderly adults and Parkinsonian patients implicates deterioration in the neural mechanisms that govern postural stability, which is evident by problems in maintaining both static and dynamic balance. Wall et al. present body-tilt information to the subject to prevent fall in the balance impaired. The precursor prosthesis is a wearable, distributed, modular design, which builds upon the single-axis research device and consists of a 3-D motion sensor array, a central processor, and vibrotactile stimulators. Kuo also reported selection of control strategies used by human in response to small perturbations to stable upright balance. A human postural optimal control model is also provided to analyze and estimate balance dynamic. There are several modes of adaptation to postural perturbations used to withstand balance disturbances and reduce the effects of muscle fatigue.

Stand and Walk

Although most of us take it for granted, walking is actually a complex task that requires intricate neural control. Successful navigation through our changing daily environments requires the ability to adapt locomotor outputs to meet a variety of situations. Jonic et al. use three supervised machine learning (ML) techniques for prediction of the activation patterns of muscles and sensory data, based on the history of sensory data, for walking assisted by FES. Popovic et al. design an instrument to improve quality of life in stroke/spinal cord injury subjects with rapid prototyping and portable FES systems. One of the main features of both stimulators is that they can reliably measure muscle EMG activity and use this signal to trigger and control the stimulation sequences.

A patient-driven control strategy for standing-up and sitting-down was experimentally tested on two paraplegic patients by applying functional electrical stimulation to the quadriceps muscle It is based on an inverse dynamic model (IDM) that predicts the stimulation pattern required to maintain the movement as it is initiated by the patient's voluntary effort.

Biofeedback Control

A major application for biofeedback is to provide tools for detecting and controlling physiological state, like heart rate, EEG and muscle activity. In the control theory view, the controller of biofeedback system is often patient's brain. The most import thing is to create a feedback path. Figure illustrates the three stages of biofeedback systems, measurement, signal processing and presentation.

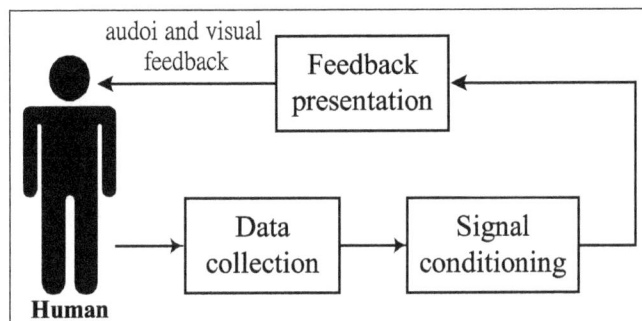

The biofeedback control system.

In 1995, Moran et al. develop a biofeedback cane system to measure axial cane force while walking. An audio alarm can be programming to sound within the adjustable limits according to cane load magnitude. Lee et al. provided a device to help improve balance and postural control by using biofeedback for standing-steadiness and weight-bearing training. Wu also presented a system that estimates and displays, in real-time, the location of the center of gravity of the human body relative to the feet was developed and then used in a biofeedback training program for improving the postural instability caused by deterioration of the Proprioc eptive system in elderly patients with significant diabetic sensory neuropathy during perturbations of a support platform.

In biofeedback, the decision-making and control computing is all accomplished by human brain. So, The effect and performance of the biofeedback is dominated in the accuracy of the measured signal, the meaning of the signal processing and the clear presentation.

Drug Delivery for Optimal Therapy

Drug can be used to regulate physiological variables, like blood pressure, blood glucose and heart rate. Due to the toxicity and side effect of drug, precise dosage regimens are important. In addition to, timely drug delivery will help patient recovering more quickly. Anesthesia is also an important drug application in surgery. The behavior of drug delivery includes two model, pharmacokinetic-pharmacodynamics model and hemodynamic model. Figure depicts pharmacokinetics and pharmacodynamics as the fundamental elements of pharmacology. In the following, we will discuss some drug delivery applications, which involve automatic technique to achieve the goal.

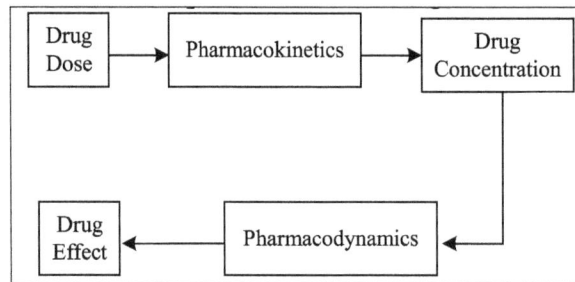

Bridge between pharmacokinetics and pharmacodynamics.

Immune System Control: HIV Example

Human immune system performs as an ingenious control system. The control objective is to resist external invader. Sometime immune system may fail, we need to deliver drug recovering original immune function. Over the last several years, most significant progress has been made in our understanding of human HIV infection. Maximal suppression of the viral load to below detectable levels has been achieved using highly active antiretroviral therapy (HAART). HAART because the high drug doses used have adverse side effects that make adherence to therapy very difficult. A regimen that could reduce dosage requirements while maintaining control over viral plasma levels might not only increase patient adherence but the overall health of the patient by reducing side effects.

Jeffrey et al. showed an application of control theory to human immunodeficiency virus (HIV) models. Brandt et al. describes a continuous differential equation model of the interaction dynamics of HIV-1 and CD4 and CD8 lymphocytes in the human body. They also demonstrate several methods of stable control of the HIV-1 population using an external feedback control term that is analogous to the introduction of a therapeutic drug regimen.

None of these immune models, however, can completely exhibit all that is observed clinically and account for the full course of the disease. The main reason for the models' limitation is lack of a good understanding of the immunology of the human body against HIV. Biological systems exhibit multicompartmental interactions that are usually not well understood and as a result, cannot be accurately modeled mathematically.

Blood Pressure Regulation

In order to reduce the blood lost during operation, a blood pressure regulation system is necessary. The main method is the infusion of sodium nitroprusside in order to lower blood pressure in patients who have undergone surgery. In tradition, the bolus injection can rapidly decrease blood pressure, but has disadvantage that the effect diminishes rapidly and it can only be applied periodically in order to avoid cyanide poisoning. Therefore, the idea of continuously controlled release of the drug has been proposed. The controlled release method has the advantage of achieving lower blood pressures over longer periods of time. The control problem is to find the correct dose, which quickly lowers the blood pressure to the desired level, while avoiding a drug overdose.

Adapting several parameters, especially in a nonlinear controller, can lead to undesired behavior because it is non-trivial to predict the closed loop system response for the whole operating range and for any possible parameter variation. Furutani et al. developed a state predictive controller to cope with the dead time existing in the responses for the drug delivery. The adaptive control is also used in mean arterial pressure through the intravenous infusion of sodium nitroprusside.

Many different controllers have been designed for the problem of blood pressure regulation and most of them adapt several parameters in order to deal with the uncertainties of the system. However, due to the fact that the ultimate goal is to design a controller that can be used in a clinical environment, the controller should be as simple as possible.

Blood Glucose Regulation

Diabetes mellitus is a disease characterized by the inability of the pancreas to regulate blood glucose concentration., such that exogenous insulin is required to control the disease. Almost all insulin-dependent diabetic subjects live with a conventional or intensified insulin therapy regimen. In order to improve the efficiency of dosing, the aim is to establish closed-loop control of blood sugar level (BSL), mimicking the endocrine pancreas. In general, a closed-loop feedback system for insulin delivery consists of a blood glucose sensor, a controller, and an infusion pump. The fig and fig display the control diagram in partially closed-loop control strategy and real-time closed loop control strategy. It is revealed that partially control needs manual intervention in glucose measurement and insulin injection. But real-time control with subcutaneous route insulin injection is fully automatic. This method is feasible in patients in intensive care.

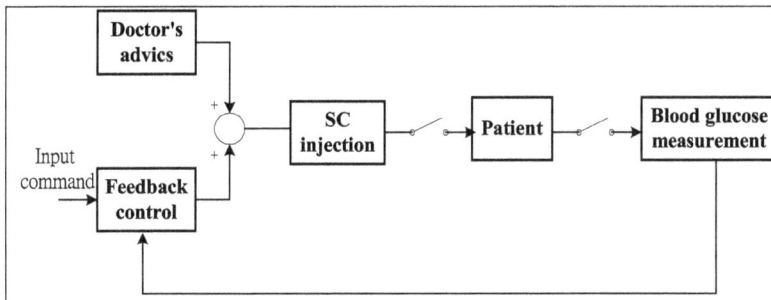

Partially closed-loop control strategy of conventional insulin therapy.

A closed-loop control system was constructed to use continuous glucose monitoring system in a real-time manner, coupled with a proportional integral control algorithm based on a sliding scale approach, for automatic intravenous infusion of insulin to patients. Neural prediction control had been used in closed loop control using subcutaneous glucose measurement and injection. The result also demonstrates that the real-time control of blood glucose is feasible.

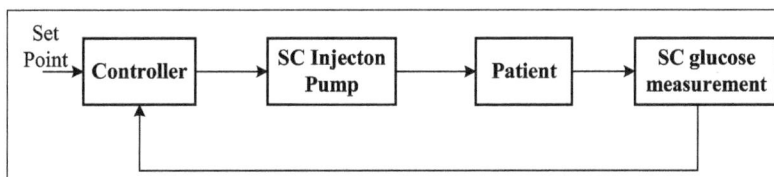

Closed-loop strategy with subcutaneous route insulin injection.

The primary need for constructing an artificial pancreas is as a reliable, long-term in vivo sensor for glucose concentration, a device that is currently unavailable. A key tenet from robust control theory is that controller performance is directly linked to model accuracy. The issues such as time delay and sensor dynamics directly affect closed-loop performance. Ongoing work is exploring methods for capturing variations in the nonlinear patient model using the linear model with variable parameters and measurement data. Fault detection is another algorithmic issue to be addressed.

Anesthesia Control

Adequate anesthesia can be defined as a reversible pharmacological state in which the patient's muscle relaxation, analgesia, and hypnosis are guaranteed. Fig shows the input/output relation of the anesthesia problem. From this figure, we can observe that anesthesia control is a complicate problem. Thus, anesthesiologists adopt the role of a feedback controller. The uses of automatic anesthesia controllers are capable of taking over and improve parts of such a complex decision process. If the routine tasks are taken over by automatic controllers, anesthesiologists are able to concentrate on critical issues that may threaten the patient's safety, the automatic controllers would be able to provide drug administration and avoid overdosing. The ultimate advantage would be a reduction in costs due to the reduced drug consumption and the shorter time spent by the patient in the postanesthesia care unit.

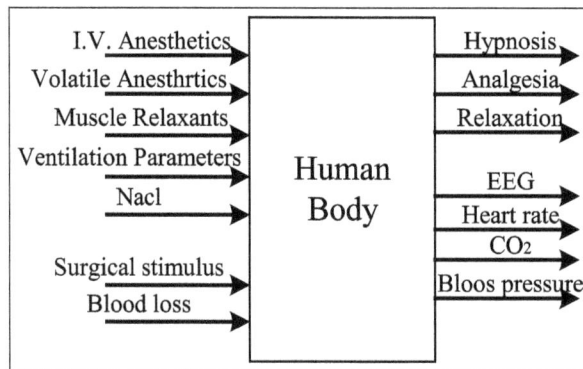

The input/output relation of the anesthesia problem.

Automatic controllers capable of regulating multiple patient outputs for higher-quality anesthesia treatment are discussed in. Maplenson et al. explore the identification problem when dealing with physiological models relating to anaesthetic drugs. The Mapleson model for drug concentration is described by algebraic equations, which are derived from the laws of physics and chemistry. Gentilini had presented a model-based closed-loop control system to regulate hypnosis with the volatile anesthetic isoflurane. Hypnosis is assessed by means of the bispectral index (BIS), a processed parameter derived from the electroencephalogram. There are also many articles assessed hypnosis by using auditory evoked potential.

The controllers cannot be used without the anesthesiologist's close supervision. The feedback systems that were proposed are SISO, whereas the goal of the anesthesiologist is to maintain several physiological variables in specified ranges. Further, the benefits of multidrug anesthesia are not yet exploited. At present, Isoflurane is used alone, and other drugs in the control schemes are not considered.

Medical Instrumentation Technique

In this topic, we will focus the work on using external device to present exogenous feedback regulation.

Temperature and Humidity Regulation

The temperature and the humidity are an important environment parameters, which not only affect our feeling, but also change our physiology. The temperature in the air could affect the lost rate of heat from skin. For this reason, many physiological parameters like body temperature, heart rate and humidity of skin vary by temperature and humidity. In medical intensive care condition, the requirement of environment parameter regulation is very crucial. The variation of temperature and humidity is conditioned by each other.

Radiant warmers and incubators are used to maintain the body temperature of newborn infants. But radiant warmers increase convective and evaporative heat loss and insensible water loss. Bouattoura et al. proposed an active humidification system to achieve high and steady humidity levels. The algorithm is based on a combination of optimal control theory and dynamic programming approach. Temperature control can also influence some physiological variables, like heart rate, breath and anesthesia levels. Qiu et al. developed temperature control system for magnetic resonance microscopy. With control of body temperature, heart rate is stabilized and repetition time during cardiac-gated studies is less variable. Thus, image quality and resolution are improved.

Accurate control is made not easily. The reason is that the level of humidity achieved is seldom related linearly to the control setting. Regulation is made even more difficult by the absence of any device for measuring humidity. The automated control system allows us more quickly to warm the subject to a target temperature while avoiding significant temperature overshoot with minimal subject deviations about the set point.

Surgery Assist Device

In this topic, we will introduce two surgery assist devices, electrosurgery scalpel and robot assist. Electrosurgery is a minimally invasive medical technique compared with conventional surgery. It consists of three functions, to destroy benign and malignant lesions, control bleeding and cut tissue. The primary cutting effect is accompanied by coagulation and hence hemostasis. The scalpel and the tissues are in ohmic contact. Then the electrical device generates radio frequency current from electrode to tissues. High heat is produced by large electrical power dissipating in ohmic contact. We have to control the power passing to tissues carefully. The task is now to determine continuously the appropriate electrical power, since this depends not only on the kind of tissue but also on the type of electrode used. Robotic assistants aid surgeons is also a minimally invasive technique. A robotic system for collaboratively or autonomously performing endoscopic procedures is the most common method. By eliminating the large incision and extensive dissection, much of the pain of recovery can also be eliminated and the length of hospital stay reduced.

Kang et al. presented the design and implementation of a new robotic system for assisting surgeons in performing minimally invasive surgical procedures. This system is designed for collaborative operation between the surgeon and the robot. A new approach in radio frequency electrosurgery, used for tissue treatment, is achieved by using a new process control method. An external control unit allows a commonly available rf-generator to automatically supply the appropriate power for differing tissue types, thus ensuring best cutting quality. The sparks, generated during the scalpel electrode interaction with the tissue, appear statistically distributed.

In terms of further technical development of surgical robot, one focuses on making the autonomous surgical procedures more robust. Careful modeling and control of the thread tension is critical in avoiding tissue tearing and pulling the needle from the stitcher or the holder.

Artificial Hearts

Cardiac assist device is used to replace the natural heart and to wait for available organ donor. It can be lowered the load of impaired heart. We need to control the artificial heart provide enough flow rate to meet the requirement of patients activity.

An automatic physiologic control system for the actively filled, alternately pumped ventricles, in long-term use had been developed by Kim et al.. The automatic control system ensure the device maintains a physiologic respond for cardiac output, compensate any nonphysiological condition and is stable, reliable and operate in high power efficiency. Maslen et al. given a brief survey of the artificial hearts reveals that its underlying premise is to develop active augmentations to the human biological function. The system identification method is used for developing heart model for the control of a cardiac ventricular assist device.

We can find two control issues in developing an adequate artificial heart:

- The power consumption of artificial heart must as low as possible, because the heat result from the dissipation of pump and electron could depress biocompatibility. Besides, the good efficiency can permit portable requirement from battery with suitable charge-to-charge cycle.

- Design a flow control system that ensures that the patient blood pressure is appropriate to his level of activity.

References

- Control-systems-introduction, control-systems: tutorialspoint.com, Retrieved 12 July, 2019

- Melby, Paul; et., al. (2002). "Robustness of Adaptation in Controlled Self-Adjusting Chaotic Systems". Fluctuation and Noise Letters. 02 (4): L285–L292. Doi:10.1142/S0219477502000919

- Bequette, B. Wayne (2003). Process Control: Modeling, Design, and Simulation. Upper Saddle River, New Jersey: Prentice Hall PTR. Pp. 165–168. ISBN 978-0-13-353640-9

- Control-systems-mathematical-models, control-systems: tutorialspoint.com, Retrieved 11 January, 2019

- Basso, Christophe (2012). "Designing Control Loops for Linear and Switching Power Supplies: A Tutorial Guide". Artech House, ISBN 978-1608075577

- Feedback-systems, systems: electronics-tutorials.ws, Retrieved 1 August, 2019

- M. Sami Fadali, Antonio Visioli, (2009) "Digital Control Engineering", Academic Press, ISBN 978-0-12-374498-2

PERMISSIONS

We would like to thank the editorial team for lending their expertise to make the book truly unique. They have played a crucial role in the development of this book. Without their invaluable contributions this book wouldn't have been possible. They have made vital efforts to compile up to date information on the varied aspects of this subject to make this book a valuable addition to the collection of many professionals and students.

This book was conceptualized with the vision of imparting up-to-date and integrated information in this field. To ensure the same, a matchless editorial board was set up. Every individual on the board went through rigorous rounds of assessment to prove their worth. After which they invested a large part of their time researching and compiling the most relevant data for our readers.

The editorial board has been involved in producing this book since its inception. They have spent rigorous hours researching and exploring the diverse topics which have resulted in the successful publishing of this book. They have passed on their knowledge of decades through this book. To expedite this challenging task, the publisher supported the team at every step. A small team of assistant editors was also appointed to further simplify the editing procedure and attain best results for the readers.

Apart from the editorial board, the designing team has also invested a significant amount of their time in understanding the subject and creating the most relevant covers. They scrutinized every image to scout for the most suitable representation of the subject and create an appropriate cover for the book.

The publishing team has been an ardent support to the editorial, designing and production team. Their endless efforts to recruit the best for this project, has resulted in the accomplishment of this book. They are a veteran in the field of academics and their pool of knowledge is as vast as their experience in printing. Their expertise and guidance has proved useful at every step. Their uncompromising quality standards have made this book an exceptional effort. Their encouragement from time to time has been an inspiration for everyone.

The publisher and the editorial board hope that this book will prove to be a valuable piece of knowledge for students, practitioners and scholars across the globe.

INDEX

www.ingramcontent.com/pod-product-compliance
Lightning Source LLC
Chambersburg PA
CBHW061252190326
41458CB00011B/3652